T0201260

The Wireless Internet of Things

The Wireless Internet of Things

A Guide to the Lower Layers

Daniel Chew

Published by
Standards Information Network

Library of Congress Cataloging-in-Publication Data:

ISBN: 9781119260578

Cover Design: Wiley
Cover Image: iStock.com/monsitj

Set in 10/12pt WarnockPro by Aptara Inc., New Delhi, India

Printed in the United States of America.

V10005052_100818

Contents

Preface

While the current state of the IoT market is fragmented across manufacturers and product lines, the future of IoT lies in interoperability. This interoperability will be enabled and communicated through easy access to technology standards developed by the IEEE and others.

This book provides an overview of several wireless standards for connectivity in the Internet of Things (IoT) and then addresses relevant wireless communications theory to help elucidate those standards. The book details the lower layers of a protocol stack in describing the wireless IoT from the bottom-up. In doing so, the book decomposes the issues to be addressed into smaller subsets.

Chapter 1 introduces the concept of the wireless internet of things, detailing background information and giving a few examples of applications. Those applications drive the requirements for the wireless links that enable the wireless IoT. The book then introduces the concept of protocol stacks, such that the wireless links can be decomposed into layers with specific responsibilities. The book gives several examples of protocol stacks and the slight differences in functional decomposition across layers. The fact that the lower layers, physical and media access, are generally covered by independent standards bodies is addressed. This book then sets its scope on those lower layers used for link establishment, channel access, error detection, and modulation. This is done by following a unified lower layer model with three layers, radio, Modem, and Media Access Control.

Chapter 2 outlines physical and media access layers for several popular open wireless standards used by IoT applications. This book focuses on open standards defined by independent standards bodies. Several popular wireless IoT protocols are referenced in a nominative sense in order to link the open standards to applications with which the reader may be more familiar. Those popular protocols are Bluetooth (formerly IEEE 802.15.1), IEEE 802.15.4, and ITU G.9959. This book focuses on low-power wireless links for the IoT; however, this chapter also briefly discusses Wi-Fi as Wi-Fi is important to many IoT applications. Wi-Fi appears again in later chapters as Wi-Fi interacts with the standards identified. This chapter can be used as a quick reference guide for the various protocols. This chapter also introduces concepts that will be explored in the later chapters, and informs the reader as to where in the book more information on that topic can be found.

Chapter 3 is dedicated to the aforementioned Radio layer. Radio front-ends are explored in this chapter. There is a slant toward software-defined radio implementations of IoT protocols, but multiple radio hardware topologies are explored. This chapter reviews the concept of link budgets and goes through examples. This chapter also addresses complex channel models employing both large and small-scale fading.

Chapter 4 focuses on the MODEM. This chapter covers the concepts of the complex-envelope signal model, modulation, demodulation, synchronization, and spread spectrum. Linear and angular modulation schemes as used in the identified open standards are explored, both in terms of background theory and also as to why a particular standard would choose a particular modulation scheme. Synchronization techniques for carrier and symbol recovery are discussed. The chapter ends with a discussion on the various spread spectrum techniques employed in the physical layers of the wireless IoT.

Chapter 5 describes the Media Access Control layer. Channel access schemes commonly employed by the wireless IoT standards, such as CSMA, are detailed. The chapter describes the different bands used by the wireless IoT standards. In particular, the 2.4 GHz Industrial, Scientific, and Medical band, and the congestion therein, is detailed. The chapter describes various interference and interference mitigation techniques employed in wireless IoT standards. The chapter concludes with a discussion on error correction and detection.

A recurring theme in this book has been that no one book can cover all of wireless system design. There are many books relevant to this course of study, on topics ranging from antenna design to symbol synchronization. This book provides background material in relevant theory, analysis of design choices, and numerous citations to aid the reader interested in learning more on a given subject. Each of the chapters in this book provides a list of references such that the interested reader can research the topic in more detail than can be covered in this limited scope.

It is the hope of the author that this book is useful to a variety of readers. Developers of platforms and applications of IoT will benefit from this book that provides a practical survey of standards relevant to the lightweight and low-cost needs of IoT platforms. This book will be useful for a variety of engineers involved in Digital Signal Processing (DSP), network implementation, and wireless communication, but could also be useful to the entrepreneur or hobbyist looking to understand the technology and develop the next big Thing.

Daniel Chew

Acknowledgments

I would like to acknowledge the people who helped make this book a reality. I would like to thank the series editors Jack Burbank and Bill Kasch for giving me the opportunity to write this book. I would like to thank Andrew Adams and Joseph Bruno for their review of my earliest material. I would like to thank Ken McKeever and Ryan Mennecke for their assistance and expertise in this field.

I also want to thank my family and friends who supported me through this process. Without them, this book could not have been possible.

I would like to dedicate this book to my wife, Lleona, and to my children, Marin, Everett, and Theodore.

About the Author

Daniel Chew is a member of the Senior Professional Staff at The Johns Hopkins University Applied Physics Laboratory and teaches in the Engineering for Professionals program at Johns Hopkins University. He received a Bachelor's of Electrical Engineering from the University of Delaware in 1998 and a Master's of Science in Electrical and Computer Engineering from Johns Hopkins University in 2008. His professional interests are in the Internet of Things, Wireless Communication Systems, Digital Signal Processing, and Software-Defined Radios.

About the Author

1

Introduction

This book began as a collection of observations and implementation experience that the author accumulated while researching wireless links used to enable the "Internet of Things" (IoT). Wireless communications engineers approach the challenge as a stack of layers, where the system has been decomposed into a stacked series of functions. Approaching the various wireless links used for IoT in this layered fashion helps cultivate an appreciation for the various standards that enable interoperability. This book will approach several standards for the wireless IoT from the layered perspective as found in a protocol stack. Organizing this book in the manner of a protocol stack will help the reader better navigate this book, and hopefully, shed some light on the purpose of the specifics within the different wireless standards that empower the IoT.

Let's begin with a question: What is the Internet of Things?

1.1 What is the Internet of Things?

The term "Internet of Things" has been around since the early 2000s [1]. This term refers to autonomous computing devices being networked together to perform various tasks. The term was coined by Kevin Ashton of the MIT Auto-ID center and was originally in reference to Radio Frequency Identifier (RFID) information being made available on the Internet [2]. RFID is a technology that allows objects to be tagged with devices that transmit identification information. RFID allows for the automatic identification and tracking of those tagged objects. This information can be sensed, gathered, parsed, and posted to the internet by way of automated and interconnected computing devices. The term "Internet of Things" has since grown to encompass far more applications and technologies than the original RFID reference.

There are a number of application areas that have either been adopted into or have grown from the Internet of Things, including:

- home automation
- medical devices

The Wireless Internet of Things: A Guide to the Lower Layers, First Edition. Daniel Chew.
© 2019 by The Institute of Electrical and Electronic Engineers, Inc. Published 2019 by John Wiley & Sons, Inc.

- industrial control
- smart grid
- distributed sensor networks
- and others

The Internet of Things is not a new concept, technology, or set of products, but is rather a natural evolution of networked computing technology, enabled primarily through affordable processing and connectivity. IoT is an extension of the "ubiquitous computing" concept popularized by Mark Weiser [3,4]. The size and cost of computing power is and has been decreasing for decades. This decrease in size and cost has resulted in small, inexpensive embedded devices, which are ideal for sensor and interface applications. Combined with the ease of connectivity provided by a robust and varied infrastructure consisting of wired, terrestrial cellular, satellite, and local wireless communication technologies, the rise of the Internet of Things is the natural consequence. While all of the technologies that comprise the Internet of Things are important, it is connectivity, particularly wireless connectivity, that is a fundamental component shaping many of the choices made in the implementation of IoT devices.

Figure 1.1 [5] illustrates the wide reach of this technology in both "vertical" and "horizontal" markets. The "vertical markets" address the needs of a specific group of consumers, and "horizontal markets" seek to address the needs of a wide group of consumers. By making use of technologies such as ubiquitous computing and wireless communications, the IoT transforms objects from being "traditional" to "smart." In Figure 1.1, these smart objects are grouped into domain-specific applications (vertical markets) while network-computing services form domain-independent services (horizontal markets).

These network-computing services are sometimes called "The Cloud." What is "The Cloud"? There is a humorous answer to that question: "There is no cloud, just someone else's computer."

"The Cloud" is a collection of computation and data storage resources made available to end-consumers by a service provider. End-consumers gain access to these resources through the Internet. This collection of computation and data storage resources is shared across the large number of end-consumers with whom the service provider has some contract.

"Cloud Computing" is where computational tasks are offloaded from local devices and executed on remote, presumably larger and more powerful, devices. The local devices make requests of the remote, more powerful, "cloud" devices. The cloud devices execute the request and provide the results to the local, smaller, devices that directly interface with the end-user.

Wireless IoT technology interfaces with "The Cloud" and "Cloud Computing" to provide many different end-user applications. For example, an end-user may use their smart phone to access a cloud data center that is updated with the status of various sensors. In that example, the wireless IoT devices form a "device

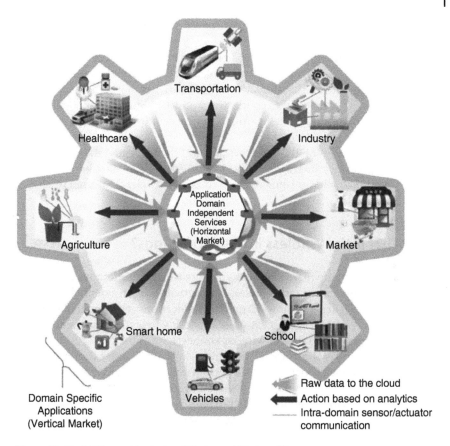

Figure 1.1 The IoT Across Vertical and Horizontal Markets [5]

network" that sends information through a gateway to a server "in the cloud." The end-user can then access that information by using a personal wireless data device to log into the repository of sensor data stored on the remote server.

While the specific implementations of each of these application areas may be quite different, they all rely on the ability to remotely monitor, manage, and actuate distributed devices. IoT technology has enabled a wide variety of applications and is already deployed across markets as disparate as health care and power grids. World market analyses have made forecasts predicting the continuing rise of IoT applications and significant contributions to the world GDP [6].

An interesting trend within the IoT is the accessibility of development to individuals, which is a by-product of low-cost processing [7]. The availability of inexpensive general purpose embedded processing devices means that device and application development is no longer limited to companies with substantial development and manufacturing budget. Hobbyists and members of the Maker community are able to make use of these platforms to create their own devices for their own unique applications.

With so much development in this field, it is clear there is a risk of fragmentation and a lack of interoperability. Without interoperability, nothing in Figure 1.1 would function. Therefore, the future of IoT lies in interoperability. It is this interoperability that makes connectivity possible. This interoperability will be enabled and communicated through easy access to technology standards developed by the IEEE and others.

This book will focus on the wireless aspects of the IoT, and the standards that enable the necessary interoperability. To that end, there must be provided a disambiguation in what is being referred to as the "wireless" Internet of Things in this book.

1.2   What is the Wireless Internet of Things?

For applications of the IoT, as networks of increasingly autonomous computing devices performing some task, wireless connectivity is often essential. Consider Figure 1.2 [8]. Figure 1.2 shows disparate applications, all connecting to the internet by way of wireless access points. Those wireless access points alone demonstrate the importance of wireless connectivity to the IoT.

Moreover, many of those application areas shown in Figure 1.2 would not be possible without locally networked devices. Wireline connectivity can establish a network of automated computing devices and connect those devices to cloud-based services. Wireless connectivity provides benefits in deployment that are unmatched by wireline solutions. Numerous sensor applications simply will not

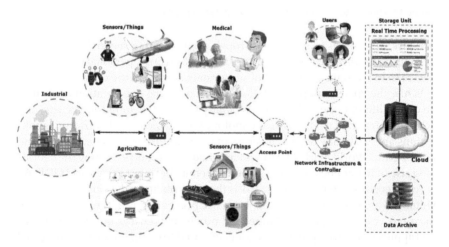

Figure 1.2 The IoT Application Areas and Wireless Connectivity [8]

function without mobility, which requires wireless connectivity. For these reasons and others, wireless connectivity is a key element to the success of IoT.

Using the term "wireless Internet of Things" narrows the conversation to focus on that wireless connectivity as opposed to cloud-based services and other aspects of popular IoT applications.

1.3 Wireless Networks

Networking is essential for the wireless IoT. Different types of networks exist to satisfy the needs of different end-user applications. Therefore, while not the focus of this book, a brief discussion of the various types of wireless IoT networks is necessary to better understand the functions of the lower layers that will be covered in the subsequent chapters.

1.3.1 Network Topologies

A network topology is the organization of nodes in a given network of nodes. A common network topology for the wireless IoT is the "star" topology [9]. The star topology is illustrated in Figure 1.3. The star topology is called such because all network traffic converges onto a single point. If any data is intended

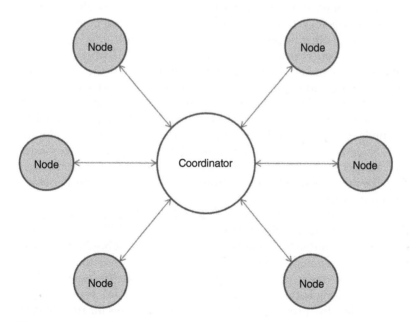

Figure 1.3 Star Topology

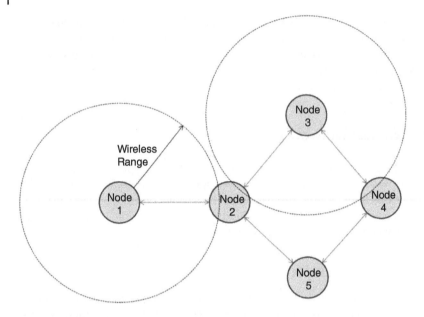

Figure 1.4 Mesh Topology

to travel from one node to another, that data must still travel through the central point of the star topology. Under the star topology, the central point serves as a coordinator for all other nodes in the wireless IoT network.

Wireless IoT networks can also be organized into "mesh" networks [9]. These mesh networks are sometimes called "peer-to-peer" networks. The mesh topology is illustrated in Figure 1.4. The nodes in the mesh network can establish links to other nodes within wireless range. Only the wireless ranges for nodes 1 and 3 are shown in in Figure 1.4 to avoid clutter. In order to propagate data from one node to another, routing must be established between the nodes. In Figure 1.4, Node 1 can only communicate with Node 2. Node 2 can see Nodes 1, 3, and 5. Node 3 can see Nodes 2 and 4. In order for Node 1 to send a message to Node 4, the message must be routed from Node 2 to either Node 3 or Node 5 and then routed again to Node 4.

Figure 1.4 shows the complexity of routing in the mesh network topology as compared to the simplicity of the star topology in Figure 1.3. The star topology requires that one coordinator node keep contact with all subordinate nodes. The mesh network topology allows for a more flexible structure to take shape, but requires routes to be in place for nodes to communicate. The exact method of establishing routing between nodes for a given wireless IoT system is specific to the wireless IoT protocol. A discussion on algorithms for routing data between nodes exceeds the scope of this one book.

1.3.2   Types of Networks

In addition to topologies, there are different types of networks designed for different scales and uses. For example, a Local Area Network (LAN) is a network intended for a single building or campus [10]. A Wide Area Network (WAN) is a network intended for an entire country or continent [10]. The Internet is an example of a WAN. A local wireless Internet connection, such as the type one might have at home with a Wi-Fi router, is referred to as a Wireless LAN (WLAN). The wireless router also provides a "gateway" for all the computing devices in your home to access the Internet. This concept is illustrated in Figure 1.5. Several personal computers and a smart phone are connected to a WLAN. The WLAN is established by a Wi-Fi router that also provides a gateway for Internet access.

A Wireless Personal Area Network (WPAN) is a network formed for data flow between the user's own personal devices. A WPAN is therefore smaller in scale than a WLAN. An example is illustrated in Figure 1.6. In this example, a wireless PAN is formed as a star topology where the central node, a smart phone, takes the role of coordinator. Figure 1.6 illustrates how a Bluetooth connection between a user's smart phone and associated peripherals such as wireless headphones is an example of a WPAN [11].

Gateways play an important part in many IoT applications. The concept is illustrated in Figure 1.7 and Figure 1.8 for the star and mesh topologies, respectively.

In Figure 1.7, a wireless "device network" connects various devices to a coordinating node in a star topology. This coordinating node then accesses a Wi-Fi router to send data to a cloud server. The Wi-Fi router serves as a "gateway."

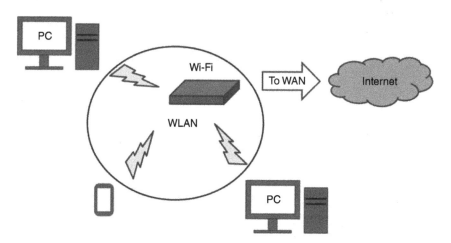

Figure 1.5 WLAN and Gateway to WAN

Figure 1.6 Wireless PAN

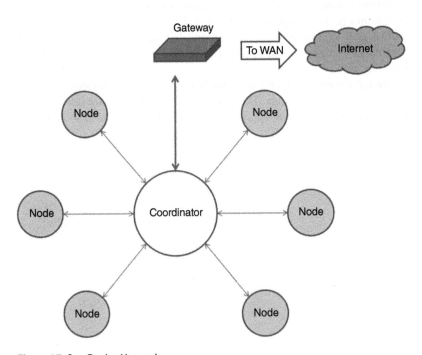

Figure 1.7 Star Device Network

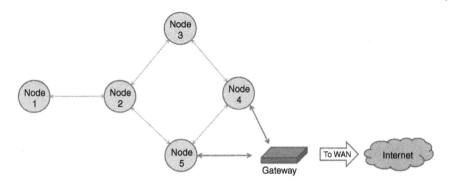

Figure 1.8 Mesh Device Network

Data from the device network is funneled to the coordinating node, which then passes the data to the gateway where it will be sent into the larger Internet.

In Figure 1.8, a number of nodes are configured in a mesh network that exists at the periphery of the larger Internet. Data from the furthest nodes must be routed and the funneled to the gateway in order to export data to the Internet.

Reference [12] provides an example use-case for a wireless device network configured in mesh topology. The use-case is illustrated in Figure 1.9. Figure 1.9 shows a wireless device (sensor) network designed to track cattle position and health. Node 1 is composed of multiple sensors and two transceivers. This node serves as the coordinator. The coordinator bridges between the device network and a WLAN. The other nodes are composed of multiple sensors and one transceiver. That one transceiver processes a low power communications protocol, ZigBee, and sends data through the network to the coordinator.

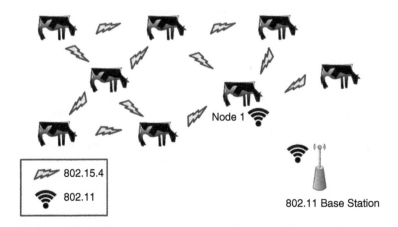

Figure 1.9 Mesh Topology of a Device Network in eAgriculture [12]

The devices in the device network are expected to operate on battery for extended periods of time. This means that the devices must be low power in order to extend battery life. The preference for low power wireless links is common in wireless sensor networks and the wireless IoT in general. This book shall focus on the standards that define the wireless links between low power devices in the wireless IoT.

1.4 What is the Role of Wireless Standards in the Internet of Things?

Wireless standards specify features for a shared wireless link such as the modulation scheme, bands of operation, and data rate, among others. Having a standard set of features in the wireless link allows for interoperability between devices created by different vendors.

Standardized wireless protocols provide a means of interoperability such that data can be exchanged between remote nodes and computing devices inside an IoT network. Wireless IoT devices conform to one of these standards and then the device can join an IoT network defined by that protocol.

The interoperability of devices utilizing these protocols rests upon conformance to a wireless standard. The standards themselves are written in such a way that a design can be tested for conformity. The standards offer little in the way of justification for the limits imposed or choices made for that standard. One of the primary goals of this book is to elucidate wireless standards pertaining to the IoT. To that end, this book approaches wireless IoT standards as "protocol stacks." A theoretical background for the wireless IoT must be provided from the ground-up. From there, the pieces can be tied together to describe in detail some of the most common wireless IoT protocols.

1.5 Protocol Stacks

A "protocol stack" is a series of layers of processing, each "stacked" upon the other. This concept is illustrated in Figure 1.10. In this example, the middle layer is labeled "Layer N" and the layers immediately above and below are labeled "Layer N + 1" and "Layer N-1," respectively. This is to demonstrate the sequential nature of the layers in data processing. Received data comes up from the physical medium and is processed sequentially by each layer from lowest to highest. The higher layers send data down to the lower layers to be transmitted across the physical medium.

Most communication systems, including the protocols for the wireless IoT, are organized into a protocol stack [13]. Such a layered architecture provides

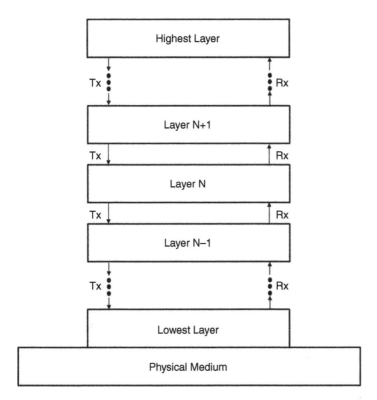

Figure 1.10 Example Protocol Stack

a natural means to decompose the various functions within a communications system.

With the basic concept of a protocol stack in mind, two of the most famous protocol stacks will now be discussed. Those are the Open Systems Interconnection reference model and the TCP/IP reference model.

1.5.1 The Open Systems Interconnection Reference Model

One of the most common protocol stacks used as a template for layered design is arguably the seven-layer Open Systems Interconnection (OSI) reference model defined by the International Standardization Organization (ISO) [14]. The seven-layer OSI model is illustrated in Figure 1.11. This stack is often used as a guide to understanding other protocol stacks.

Each layer in the seven layer OSI reference model has a specific function:

- Application: This is the process that is ultimately producing and consuming data.

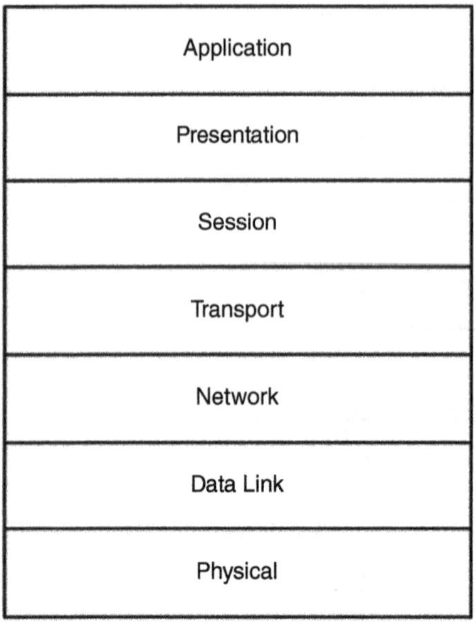

Application
Presentation
Session
Transport
Network
Data Link
Physical

Figure 1.11 Seven-Layer OSI Reference Model

- Presentation: Provides independence to application processes by structuring the data.
- Session: Provides control and synchronization between application processes communicating across the network (e.g., starting a "session," ending a "session").
- Transport: Packetizes the data, sequences the data, and handles connection-oriented or connectionless delivery.
- Network: Routes the data across the network.
- Data Link: Controls access to the physical medium. Corrects for errors in received data.
- Physical: Transmits and receives data across the physical medium.

1.5.2   The TCP/IP Reference Model

The TCP/IP reference model is the stack upon which the Internet rests. While the OSI reference model is popular for academic study or a common point of reference to compare other protocol stacks, it has not gained popularity in actual implementation. The OSI reference model and the "TCP/IP" reference model were competing protocol stacks for what was to become the Internet [15,16]. To summarize the result of years of publications and debate, the TCP/IP model won.

The TCP/IP reference model is illustrated in Figure 1.12 [17]. The corresponding layers in System A and System B interact by way of transmitting data through the lower layers and processing data received from the lower layers.

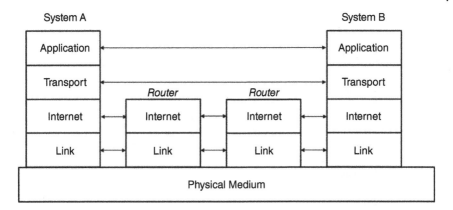

Figure 1.12 TCP/IP with Routers

Routers in the network between System A and System B are represented by the two partial stacks. The routers do not need to parse the whole stack, only enough to route packets between System A and System B.

Each layer in the TCP/IP four-layer stack has a specific function:

- Application: This is the process that is ultimately producing and consuming data.
- Transport: Packetizes the data, sequences the data, and handles connection-oriented or connectionless delivery.
- Internet: Routes the data across the network.
- Link: Controls access to the physical medium. Corrects for errors in received data. Transmits and receives data across the physical medium.

The layers of the TCP/IP reference model and the layers of the OSI reference model can be mapped as shown in Figure 1.13. As can be seen, the TCP/IP reference model is much simpler than the OSI reference model. Functions placed by the OSI reference model into the Presentation and Session layers are left solely to the Application layer in the TCP/IP reference model.

Figure 1.13 demonstrates that there are functions common between ostensibly different designs of protocol stacks. These stacks can often be mapped or translated between one another. The fact that mapping is often possible provides a mechanism such that concepts common to different stacks for different applications can be discussed in the abstract.

1.5.3 The IEEE 802 Reference Model

IEEE has published a family of standards for personal, local, and metropolitan area networks. These are the IEEE 802 set of standards. The protocol stack of

Application	Application
Presentation	
Session	
Transport	Transport
Network	Internet
Data Link	Link
Physical	

Figure 1.13 Mapping TCP/IP to OSI

the reference model for these standards is derived from the OSI model, and maps to the OSI model as shown in Figure 1.14. The IEEE 802 standards only define the physical and data link layer. The data link layer is broken into two sub layers, the Media Access sublayer and the logical link control sublayer. The definitions of the upper layers are left to other standards.

The Logical Link Control layer (LLC) for all IEEE 802 standards is defined in IEEE 802.2. The relationship between these standards is shown in Figure 1.15. Only Ethernet (802.3), 802.15.4, 802.15.1 (formerly Bluetooth), and Wi-Fi (802.11) are shown in Figure 1.15 for the sake of brevity. These various protocols share a common definition for LLC layer, that being IEEE 802.2.

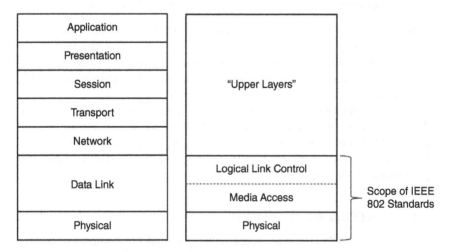

Figure 1.14 The IEEE 802 Reference Model

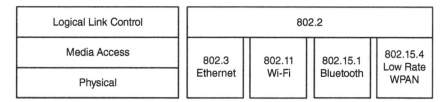

Logical Link Control	802.2			
Media Access	802.3 Ethernet	802.11 Wi-Fi	802.15.1 Bluetooth	802.15.4 Low Rate WPAN
Physical				

Figure 1.15 The Relationship Between IEEE 802 Standards

The physical (PHY) and media access control (MAC) layers are defined for each different standard. Therefore, the PHY and MAC deserve some special attention.

1.5.4 A Layered Model for IoT Operations

Layered models can be used to describe systems other than wireless links. Numerous articles and papers have employed layered models to describe IoT operations in the abstract. Such a layered model encompasses all of the IoT, from the small sensing device through the wireless protocol to the cloud operations. There are several different layered models employed in literature to describe the operation of the IoT. One of the most common of these is the "three-layer model" [18–20]. The "three-layer model" shall be used here for its simplicity. This model is illustrated in Figure 1.16.

The "Perception" layer is called such because this is the layer where information is gathered. This layer handles any sensor that is attached to the wireless IoT links. The data gathered could be from any source relevant to the application. The sensors could be any number of things from vibration sensors for a security system to sensors used to gather data on soil conditions for agricultural applications.

The "Network" layer handles all transportation of data. This is the layer that encapsulates much of what this book will discuss. The operation of the wireless links between sensors and network coordinators are encapsulated here. Routing between the network nodes is encapsulated here. Funneling information to a gateway and sending that information to remote servers is encapsulated here.

Application
Network
Perception

Figure 1.16 Three-Layer Model for Wireless Sensor Networks

The "Application" layer represents the desired application. This layer handles the user interface and decisions made at the top of the stack. The application is the layer that might exist in "the cloud."

The operation of the wireless link that empowers the device network is encapsulated in the "Network" layer of the model in Figure 1.16. This book will focus on that wireless interoperability.

1.6 Introduction to the Protocols for the Wireless Internet of Things

This book will delve into concepts that help define the lower layers of wireless IoT. Specific protocols that will be explored will include:

- IEEE 802.15.4,
- Bluetooth (previously standardized as IEEE 802.15.1),
- and ITU-T G.9959.

The IEEE 802.15.4 provides standards for PHY and MAC layers for low-rate wireless personal area networks. A subset of the PHY and MAC layers defined therein are used by wireless IoT protocols.

The Bluetooth standard is maintained by the Bluetooth Special Interest Group (SIG). The lower layers of Bluetooth, PHY, and MAC were once standardized as IEEE 802.15.1. Now those lower layers are maintained by the Bluetooth SIG.

ITU-T G.9959 provides a standard for the PHY and MAC layers for short range narrow-band digital radio communication transceivers. The Z-Wave protocol for the wireless IoT uses the ITU-T G.9959 standard for its lower layers. The upper layers are maintained by the Z-Wave Alliance.

Each protocol has a stack, and there are many features common to these stacks. One common theme is that the lower two layers, PHY and MAC, are maintained by neutral standards bodies whereas the upper layers are maintained by an industry group. The details of the upper layers of these wireless IoT protocols may not be free for a developer to use. A developer must check with the specific industry alliance before publishing information on those upper layers or developing solutions using those upper layers.

Wi-Fi, which follows the IEEE 802.11 standards, will also be discussed. This book focuses on the standards for low power wireless IoT protocols. Wi-Fi does not fit that criteria; therefore, a detailed discussion on Wi-Fi is beyond the scope of this book. While the focus of this book is on IoT devices and the wireless protocols between them, Wi-Fi plays an important role in the wireless IoT and some discussion is necessary.

1.7   The Approach of this Book

Chapter 2 will provide background and operational information on the protocols identified in Section 1.6. The remainder of this book will follow a layered approach to providing an in-depth analysis of the lower layers of the wireless IoT standards. To accomplish this, a hybrid model of IoT protocols stacks will be followed charting a path through the lower layers of the protocol stacks.

As will be shown in Chapter 2, the protocol stacks for the wireless IoT can be simplified to fewer layers than that shown in Section 1.5. This hypothetical stack is illustrated in Figure 1.17. There are only two layers that will be covered in this book: Physical and MAC.

The physical layer is split into two hypothetical sublayers, "Radio" and "MODEM." These two sublayers will be covered in Chapters 3 and 4, respectively.

The Radio layer will be defined as encapsulating the physical interface to the spectrum. The Radio layer chapter will cover the theory and technology behind the wireless links described in the wireless standards of interest. This will include topics such as radio hardware and channel effects.

The Modem layer will be defined as encapsulating the modulation and demodulation algorithms necessary to convert bits into waveforms and vice versa. The Modem layer chapter will cover the theory and algorithms behind the wireless standards of interest. This will include topics such as modulation, demodulation, and spread spectrum.

Chapter 5 will focus on the MAC layer. The "Media Access Control layer" (MAC) is a common name for the layer that manages and controls access to limited Physical layer resources. The OSI model considers MAC to be a "sublayer" of the larger "Data Link" layer. The standards defined under IEEE 802 also identify a MAC sublayer. Within the IEEE 802 literature, the MAC "sublayer" is sometimes referred to simply as the MAC "layer." Some other protocols, such as Z-Wave, specifically delineate a MAC layer within the standard.

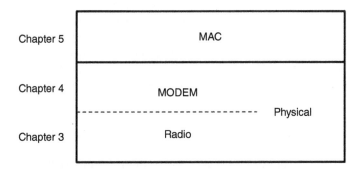

Figure 1.17 Simplified Protocol Stack

Upper layers must send data to the MAC, which then manages transmissions parameters. The MAC manages the reception of data and provides complete packets to upper layers. The MAC coordinates access with other nodes in the system. Spread spectrum techniques are typically discussed as physical layer properties. However, spread spectrum techniques do require some negotiation and coordination at higher layers.

The MAC layer is important to the wireless IoT as it controls access to the physical medium in contentious environments. This chapter will delve into the background theory necessary to understand the operation of media access control and the wireless standards specifying operations thereof. Topics will include multiple access techniques and error correction.

References

1 F. Mattern and C. Floerkemeier, "From the Internet of Computers to the Internet of Things," in *From Active Data Management to Event-Based Systems and More*. Berlin, Heidelberg: Springer, 2010, pp. 242–259.
2 C. R. Schoenberger. (2002, Mar. 18). The internet of things. *Forbes* [Online]. Available: https://www.ieee.org/content/dam/ieee-org/ieee/web/org/conferences/style_references_manual.pdf
3 M. Weiser, "The computer for the 21st century," *Sci. Am.*, vol. 265, no. 9, pp. 66–75, 1991.
4 M. Weiser, R. Gold, and J. S. Brown, "The origins of ubiquitous computing research at PARC in the late 1980s," *IBM Syst. J.*, vol. 38, no. 4, pp. 693–696, 1999.
5 A. Al-Fuqaha, M. Guizani, M. Mohammadi, M. Aledhari, and M. Ayyash, "Internet of Things: A survey on enabling technologies, protocols and applications," *IEEE Commun. Surveys Tuts.*, vol. 17, no. 4, pp. 2347–2376, 2015.
6 L. Yang, C. Yao, T. Nguyen, S. Gurumani, K. Rupnow, and D. Chen, "System-level design solutions: Enabling the IoT explosion," in *2015 IEEE 11th Int. Conf. ASIC (ASICON)*, Chengdu, China, Nov. 2015, pp. 1–4.
7 K. J. Singh and D. S. Kapoor, "Create your own Internet of Things: A survey of IoT platforms," *IEEE Consum. Electron. Mag.*, vol. 6, no. 2, pp. 57–68, 2017.
8 L. Farhan, S. T. Shukur, A. E. Alissa, M. Alrweg, U. Raza, and R. Kharel, "A survey on the challenge and opportunities of the Internet of Things (IoT)," in *2017 IEEE 11th Int. Conf. Sens. Technol. (ICST)*, Sydney, Australia, Dec. 2017, pp. 1–5.
9 J. Misic and V. Misic, *Wireless Personal Area Networks: Performance, Interconnections and Security with IEEE 802.15.4*. West Sussex: John Wiley & Sons Ltd., 2008.
10 A. S. Tanenbaum, *Computer Networks*. Upper Saddle River: Prentice Hall, 2003.

11 P. Johansson, M. Kazantz, and M. Gerla, "Bluetooth: An enabler for personal area networking," *IEEE Netw.*, vol. 15, no. 5, pp. 28–37, 2001.

12 P. K. M. Nkwari, S. Rimer, B. S. Paul, and H. Ferreira, "Heterogeneous wireless network based on Wi-Fi and ZigBee for cattle monitoring," in *IEEE IST-Africa Conf.*, Lilongwe, Malawi, May 2015, pp. 1–9.

13 M. R. Palattella, M. Accettura, X. Vilajosana, T. Watteyne, L. A. Grieco, G. Boggia, and M. Dohler, "Standardized protocol stack for the Internet of (important) Things," *IEEE Commun. Surveys Tuts.*, vol. 15, no. 3, pp. 1389–1406, 2013.

14 H. Zimmerman, "OSI reference model—The ISO model of architecture for open systems interconnection," *IEEE Trans. Commun.*, vol. 28, no. 4, pp. 425–432, 1980.

15 A. L. Russell, "The internet that wasn't," *IEEE Spectr.*, vol. 50, no. 8, pp. 39–43, 2013.

16 D. Meyer and G. Zobrist, "TCP/IP versus OSI," *IEEE Potentials*, vol. 9, no. 1, pp. 16–19, 1990.

17 R. Braden, "RFC1122: Requirements for Internet hosts – communication layers," The Internet Engineering Task Force, 1989.

18 L. Dan, S. Jianmei, Y. Yang, and X. Jianqiu, "Precise agricultural greenhouses based on the IoT and fuzzy control," in *IEEE Int. Conf. Intell. Transp. Big Data Smart City (ICITBS)*, Changsha, China, Dec. 2016, pp. 580–583.

19 M. Wu, T.-J. Lu, F.-Y. Ling, J. Sun, and H.-Y. Du, "Research on the architecture of Internet of Things," in *IEEE 3rd Int. Conf. Adv. Comput. Theory Eng. (ICACTE)*, Chengdu, China, Aug. 2010, pp. V5-484–V5-487.

20 M. Frustaci, P. Pace, and G. Aloi, "Securing the IoT world: Issues and perspectives," in *IEEE Conf. Stan. for Commun. Netw. (CSCN)*, Helsinki, Finland, Sep. 2017, pp. 246–251.

2

Protocols of the Wireless Internet of Things

There are many wireless standards covering the applications of the Internet of Things (IoT). Why should this be? Why not standardize to one-size-fits-all?

As the number of applications for IoT grows, new opportunities for wireless technology emerge. Each of these opportunities provides a niche for a wireless link. New wireless IoT protocols are developed to satisfy that emerging need.

There are efforts to standardize these wireless protocols. Industry groups that organize to develop components for different wireless IoT protocols often coordinate with independent standards bodies, such as the IEEE, to open their protocol standard to new developers and so that the wireless standard enjoys the benefits of being maintained by a neutral party.

Many of the standards covered in this book began with a single vendor and a specific application in mind. For example, Bluetooth was originally developed and put forward by Ericsson to provide a means for wireless connectivity between mobile phones and ancillary devices [1]. That is a very specific application space. There was an implementation in mind before the standards body ever met to create IEEE 802.15.1 and make this idea an open standard. As will be seen, this is not an uncommon occurrence. Opening the standard allows for more manufacturers to participate in the market and that participation will grow the market for that application. As will be seen, standards grow and develop new modes over time to accommodate new requirements of new markets.

This book will focus on the lower layers of several wireless protocols for the Internet of Things (IoT). Those protocols are:

- Bluetooth (formerly IEEE 802.15.1),
- IEEE 802.15.4,
- and ITU G.9959.

These are common wireless IoT protocols where each addresses a growing need in IoT applications. These protocols focus on wireless devices for Wireless Personal Area Networks (WPAN) and Wireless Sensor Networks. This chapter

The Wireless Internet of Things: A Guide to the Lower Layers, First Edition. Daniel Chew.

will provide a background on these protocols and the Wireless Local Area Network (WLAN) standard IEEE 802.11, often called Wi-Fi, as that WLAN protocol interacts with the aforementioned wireless device protocols.

2.1 Bluetooth

In 1994, the L.M. Ericsson Company wanted to develop a new wireless link as a means to provide wireless connectivity between Ericsson mobile phones and ancillary devices [1]. At the time, IEEE did not yet have a standard for Personal Area Networks. Ericsson teamed up with IBM, Intel, Nokia, and Toshiba to create the Bluetooth Special Interest Group (SIG). The Bluetooth SIG set out to develop the desired wireless standard. In 1999, the Bluetooth SIG issued the specification for Bluetooth v1.0. In 2002, the IEEE 802.15 committee approved the Wireless Personal Area Network (WPAN) standard 802.15.1 based on Bluetooth v1.0 [2]. The standard became IEEE 802.15.1-2002. The protocol stack, as illustrated in the standard IEEE 802.15.1-2002, is shown in Figure 2.1. IEEE 802.15.1-2002 provided the standard for the physical and media access layer, IEEE 802.2 provided a standard for Logical Link Control, and the higher layers were specified by the Bluetooth SIG. Much like other wireless IoT protocols, the physical and media access layers were defined by the open standard. This phase in the development of Bluetooth standards is called "Bluetooth Classic."

As shown in Figure 2.1, the Bluetooth stack has multiple Service Access Points (SAPs) between the layers defined by IEEE 802.15.1-2002 and the higher layers. The Bluetooth lower layers were designed to service a number of applications. The higher layer functions to which the various SAPs provide data are largely beyond the scope of this book.

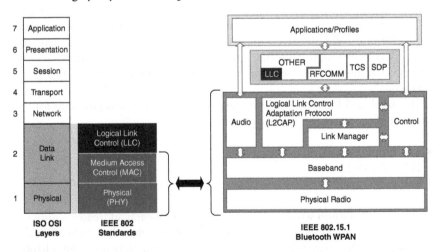

Figure 2.1 Bluetooth v1.0 Protocol Stack [2]

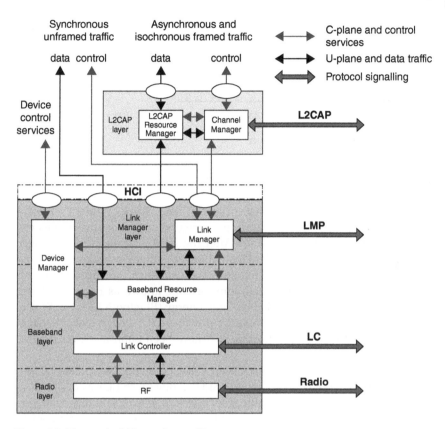

Figure 2.2 Bluetooth v1.0 Lower Layers [2]

Figure 2.2 shows that the three lowest layers are the radio layer, the baseband layer, and the link manager layer. Those layers map to layers 1 and 2 in the Open Systems Interconnection (OSI) stack.

- The radio layer is the physical layer, specifying transceiver and modulation requirements.
- The baseband layer handles forward error correction, cyclic redundancy checks, and automatic repeat requests.
- The link manager layer handles the management of the wireless link including power control, connection establishment and management, authentication, and other management functions.

As can be seen in Figure 2.2, the lower layers are abstracted from the upper layers through the Host Controller Interface (HCI). This is in reference to a common Bluetooth device architecture, shown in Figure 2.3. In the

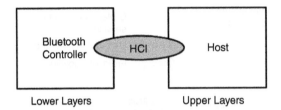

Figure 2.3 Bluetooth Device Architecture

architecture, there is a Bluetooth Controller, which serves as the radio function, and a Host, which handles upper layer processing.

Bluetooth has evolved since that inception. IEEE 802.15.1-2002 eventually became IEEE 802.15.1-2005, adding improvements such as adaptive frequency-hopping spread spectrum (AFH) [3]. After the 2005 revision, IEEE no longer maintains the standard. The task of maintaining the Bluetooth standard is now carried by the Bluetooth SIG. Bluetooth 2.0 provided an Enhanced Data Rate (EDR). The original Bluetooth definition became known as Bluetooth Basic Rate (BR). Bluetooth 4.0 provided the Low Energy (Bluetooth LE or BLE) version of the Bluetooth protocol. Table 2.1 provides a brief overview of the versions of Bluetooth Core Specification that have been published over the last two decades [4].

Bluetooth LE is of particular interest for a discussion on the wireless IoT [5]. Bluetooth LE is marketed as "Bluetooth Smart." Bluetooth LE was designed to conserve energy as would be required for wireless sensors and other devices. The Bluetooth LE protocol stack is slightly different from the standard stack. The lower layers are shown in Figure 2.4. The physical layer takes the role of the radio and baseband layers from Figure 2.2. The link layer takes the role of the link manager layer from Figure 2.2. The HCI continues to provide abstraction between the upper and lower layers of the stack.

Table 2.1 Bluetooth Versions [4]

Version	Notes
1.1	"Classic"
2	Added Enhanced Data Rate (EDR)
3	Added AMP (alternative MAC/PHY). Enabled high speed transfers through WiFi (802,11)
4	Introduced Bluetooth Low Energy (BLE)
5	Enhanced BLE

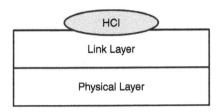

Figure 2.4 Bluetooth LE Lower Layers

2.1.1 Transceiver

Bluetooth identifies different "classes" of Bluetooth transmitters. These classes impose limits on the transmit power. Table 2.2 lists the power classes.

The transmit power for Bluetooth LE devices ranges from 0.01 mW to 100 mW, as shown in Table 2.3.

The Bluetooth standard specifies a reference receiver sensitivity of −70 dBm. The receiver sensitivity is defined by the standards as the signal strength (in power) at the receiver. The standards require that at this minimum, the receiver be able to achieve a Bit Error Rate (BER) of one error in 1000 bits.

Bluetooth devices can dynamically change the power, and may be commanded to do so.

These values can be used to calculate a link budget and link margin. Link budgets and margin are discussed in Chapter 3.

Table 2.2 Bluetooth Power Classes [4]

Power class	Maximum transmit power	Minimum transmit power
1	100 mW	1 mW
2	2.5 mW	0.25 mW
3	1 mW	N/A

Table 2.3 Bluetooth LE Power Classes [4]

Power class	Maximum transmit power	Minimum transmit power
1	100 mW	1 mW
1.5	10 mW	0.01 mW
2	2.5 mW	0.01 mW
3	1 mW	0.01 mW

2.1.2 Frequency Channels

Bluetooth operates exclusively in the 2.4 GHz industrial, scientific, and medical band. This band is internationally designated for unlicensed operations. This band is discussed in Chapter 5.

Bluetooth BR/EDR uses a channel plan of 79 frequency channels spaced 1 MHz apart. Bluetooth LE uses 40 frequency channels spaced 2 MHz apart.

Bluetooth uses frequency hopping across these defined frequency channels. The channelization of Bluetooth is, therefore, not the frequency channel plan but rather the hop plan. Frequency hopping spread spectrum techniques are covered in Chapters 4 and 5.

2.1.3 Typical Range

The operational range of a wireless system depends on many factors. Chapter 3 will review link budgets, which are necessary to calculate operational range in a given environment. As is shown in Chapter 3, the range depends upon transmitter power. Transmitter power, as shown in Table 2.2, is variable and depends on the "class" of the Bluetooth device. With that caveat in place, Bluetooth offers typical ranges between 1 meter and 100 meters [6]. These ranges are the intended operational ranges.

Bluetooth has added a feature to BLE in version 5.0 Core Specifications called "LE Coded." This adds forward error correction to BLE, which had only used cyclic redundancy checks up to that point. The addition of forward error correction reduces the data rate but may extend range due to improved immunity to bit errors. One cannot get exact values for range extension without analyzing the environment in which the system is to be used. The addition will improve range at the expense of data rate in many environments.

2.1.4 Access and Spread Spectrum

Bluetooth utilizes frequency hopping for spread spectrum. Frequency hopping spread spectrum techniques are covered in Chapters 4 and 5. Time-division duplexing is used for two-way communication in Bluetooth BR/EDR. Time-division duplexing is covered in Chapter 5.

BLE changes the paradigm. BLE is time-division duplexed between two nodes in a connection; however, BLE slave nodes do not share a physical channel. This is covered more in Section 2.1.7.

2.1.5 Modulation and Data Rate

Modulation schemes used in Bluetooth are shown in Table 2.4.

- GFSK stands for Gaussian Frequency-Shift Keying and is discussed in Chapter 4.
  - The bandwidth–time product is 0.5 for all cases.

Table 2.4 Bluetooth Modulation Types

Version	Modulation type	Data rate
Basic rate	GFSK	1 Mbps
Enhanced Data Rate	GFSK, π/4-DQPSK, 8DPSK	3 Mbps
Low Energy	GMSK	1 Mbps
Low Energy (optional)	GMSK	2 Mbps

o GMSK stands for Gaussian Minimum-Shift Keying and is discussed in Chapter 4.
- π/4-DQPSK stands for "pi-over-four" Differential Quadrature Phase-Shift Keying and is discussed in detail in Chapter 4.
- 8DPSK stands for 8-ary Differential Phase-Shift Keying and is discussed in Chapter 4.

GFSK frequency deviations are provided in Table 2.5. Bluetooth BR and EDR use the same GFSK modulation index. Bluetooth EDR only uses GFSK for the beginning of the packet, switching the modulation to π/4-DQPSK or 8DPSK for the payload data. Bluetooth LE uses a modulation index of 0.5, resulting in a GMSK modulation scheme with a wide tolerance on the modulation index implemented by a Bluetooth LE transmitter. The data rate for Bluetooth LE was originally 1 Mbps. Bluetooth Core Specification 5.0 [4] added an optional data rate for 2 Mbps. This is reflected in the doubling of the frequency deviation.

Bluetooth also imposes a minimum frequency deviation. The peak frequency deviation for transmitting a symbol must never fall below the values in Table 2.6.

2.1.6   Error Detection and Correction

Bluetooth BR and EDR use forward error correction in the form of convolutional codes. Forward error correction is discussed in Chapter 5.

Bluetooth LE originally only used cyclic redundancy checks as it is with many wireless IoT protocols. The prevalence of this is also discussed in Chapter 5.

Table 2.5 Bluetooth Modulation Indices [4]

Version	Maximum frequency deviation	Modulation index
Basic rate, Enhanced Data Rate	157.5 kHz	$0.315 \pm 11\%$
Low Energy	250 kHz	$0.5 \pm 10\%$
Low Energy (optional)	500 kHz	$0.5 \pm 10\%$

Table 2.6 Bluetooth Minimum Frequency Deviation [4]

Version	Minimum frequency deviation
Basic rate, Enhanced Data Rate	115 kHz
Low Energy	185 kHz
Low Energy (optional)	370 kHz

Optional forward error correction was introduced for Bluetooth LE in version 5.0 of the Bluetooth Core Standards [4]. This has the capacity to increase the effective range at the expense of the data rate.

2.1.7 Network Topology

Bluetooth operates on a star network topology the standard calls piconets and scatternets. A piconet may contain up to eight devices. A scatternet is created from multiple piconets. Figure 2.5 comes from [3] and illustrates this concept. Figure 2.5a shows "single slave" operation. Figure 2.5b shows "multislave" operation. Both "single slave" and "multislave" are piconets. Figure 2.5c shows scatternet operation. A master of one piconet may be a slave in another piconet to another master. Also shown in Figure 2.5c is a slave node that responds in two piconets and therefore has two master nodes. The slave devices listen to broadcasts from the master device.

This paradigm changed slightly with Bluetooth Low Energy. The slave devices advertise to the master device. This is because slave devices are considered power constrained, and therefore the slave devices spend most of their time in sleep mode. The master device is assumed to be not as power constrained and therefore listens and establishes connections [4]. A master node may establish connections with multiple slaves, but those are considered separate channels.

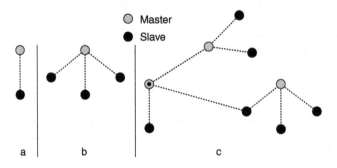

Figure 2.5 Bluetooth Topology [3]

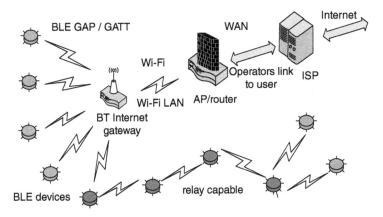

Figure 2.6 Bluetooth Low Energy Mesh Topology [8]

Bluetooth released a standard to cover mesh network topologies in Bluetooth in 2017 [7]. This standard had been in development for a few years. Reference [8] provides an illustration as to how it could be used, and that is shown in Figure 2.6.

2.2 ITU G.9959

ITU G.9959 is the International Telecommunication Union (ITU) standard covering "short range narrow-band digital radio communication transceivers." ITU G.9959 began in 2012 [9]. It was created from the lower layers of the Z-Wave protocol. The Z-Wave protocol had been proprietary. Creating the independent standard allowed interoperability with other hardware.

ITU G.9959 was originally released in 2012 to cover PHY and MAC layer specifications. In 2015, ITU G.9959 was updated to include two additional layers in the stack: Segmentation and Reassembly (SAR) and Logical Link Control (LLC) [10]. This change expanded the coverage of ITU G.9959 standard such that it now encompassed all functions of the OSI Data Link Layer. The LLC layer is a sublayer of the Data Link Layer in the OSI stack, and it allows for multiplexing and de-multiplexing to multiple and concurrent network layer implementations. This allows technology based on the ITU G.9959 standard to choose different network layer implementations, such as "IPv6 over Low-Power Wireless Personal Area Networks" (6LoWPAN) [11]. The SAR layer is one of several optional adaptation layers between the MAC and LLC layers. The scope of this book will extend only to the MAC layer.

It is important to note that ITU G.9959 itself is not a wireless IoT implementation or technology. It is a standard to which technologies conform.

Table 2.7 ITU G.9959 Minimum Sensitivity

Data rate	Minimum receiver sensitivity
19.2 kbps (R1)	−95 dBm
40 kbps (R2)	−92 dBm
100 kbps (R3)	−89 dBm

2.2.1 Transceiver

ITU G.9959 allows the transmitter to transmit at the maximum power allowed by the specific region in the band of operation. For ITU G.9959, this output power level is referred to as "nominal."

Table 2.7 shows the minimum sensitivity for ITU G.9959 receivers. At this minimum sensitivity, the Communications Error Rate (CER) is expected to be less than 10%.

These values can be used to calculate a link budget and link margin. Link budgets and margin are discussed in Chapter 3.

2.2.2 Frequency Channels

The ITU G.9959 protocol operates in sub-1 GHz unlicensed bands. Sub-1 GHz bands are not standardized internationally. The exact location of the sub-1 GHz unlicensed band depends upon the region. For example, there is the 900 MHz (Americas and Australia) Industrial, Scientific, and Medical (ISM) band and the 800 MHz (Europe) Short Range Device (SRD) band. Choosing to operate in these bands has some advantages. ITU G.9959 avoids the congestion of the popular 2.4 GHz ISM band and enjoys better propagation through the home. However, the frequencies used do vary by region; therefore, ITU G.9959 products are specific to individual regions.

2.2.3 Typical Range

The operational range of a wireless system depends on many factors. Chapter 3 will review link budgets, which are necessary to calculate operational range in a given environment. With that caveat in place, the typical operating range can be discussed. Reference [12] lists typical operating ranges for various wireless IoT protocols. Reference [12] specifically refers to Z-Wave and not to ITU G.9959; however, it is ITU G.9959 that dictates the transceiver. Therefore, based on [12], ITU G.9959 offers typical ranges of 30 meters when operating indoors and 100 meters when operating outdoors.

2.2.4   Network Topology

The 2012 version of the ITU G.9959 standard specifically stated that ITU G.9959 operates on a mesh topology and specifically referred to home automation use in the open standard. The 2015 version of the ITU G.9959 standard became more general and did not specify the mesh topology. Both versions contain an appendix A that is dedicated to Z-Wave. In that appendix, it is stated that Z-Wave operates on a mesh network topology.

2.2.5   Access and Spread Spectrum

ITU G.9959 does not employ any spread spectrum techniques. ITU G.9959 uses Carrier-Sense Multiple Access (CSMA) to mitigate contention for the frequency channel of operation. CSMA is discussed in detail in Chapter 5.

2.2.6   Modulation and Data Rate

The ITU G.9959 standard specifies three modulation rates: R1, R2, and R3. These modulation rates are tied to three frequency channels. The center frequencies of these channels are determined by region, due to the regional definition of sub-1 GHz ISM bands. Not every region offers all three channels.

ITU G.9959 uses three different data rates and specifies a different one for each, as shown in Table 2.8. The FSK schemes for R1 and R2 rates do not use any baseband filtering. R1 is Manchester encoded. The FSK scheme for R3 uses Gaussian pulse shaping with a bandwidth time product of $BT = 0.6$. FSK and GFSK are discussed in Chapter 5. Frequency deviation is $\frac{1}{2}$ the total "frequency separation."

2.2.7   Error Detection and Correction

ITU G.9959 uses no forward error correction technique. ITU G.9959 employs checksums or cyclic redundancy checks to detect whether a bit error has occurred in a received packet. Because there is no forward correction, a single bit error will require retransmission of the packet.

Table 2.8 ITU G.9959 Modulation Indices

Data rate	Frequency deviation	Modulation index
19.2 kbps (R1)	20 kHz ± 10%	1.0415
40 kbps (R2)	20 kHz ± 10%	1
100 kbps (R3)	29 kHz ± 10%	0.58

Checksums, cyclic redundancy checks, and automatic repeat requests are discussed in Chapter 5.

2.3 Z-Wave

Z-Wave was created by the ZenSys Corporation as a wireless link to support smart home products, e.g., lighting controls, thermostats, and garage door openers [13, 14]. ZenSys began marketing Z-Wave in 2003. In 2005, other companies joined ZenSys to form the Z-Wave Alliance. In 2008, ZenSys was acquired by Sigma Designs.

Z-Wave was specifically developed as a means to enable home automation products and applications. The Z-Wave physical layer focused on being low power, preserving battery life, and propagation through an indoor home environment.

The protocol stack for Z-Wave is shown in Figure 2.7 [12, 15]. There are five layers: physical (PHY), media access (MAC), transfer, routing, and application.

- The physical layer handles radio and modem functions.
- The media access layer handles Carrier Sense Multiple Access of the medium.
- The transfer layer handles error checking, acknowledgments, and repeat requests.
- The routing layer handles routing through the mesh network.
- The application layer is the seat of the application.

Reference [16] provides the layout of the Z-Wave packet across the various layers. That layout is shown in Figure 2.8. The physical layer is represented as a transmitter/antenna. The MAC layer controls framing and carrier sensing. The "Transfer + Routing" layer handles addressing and error detection.

The lower two layers of Z-Wave were standardized by the International Telecommunication Union as ITU G.9959 in 2012 [9]. That the lower layers are

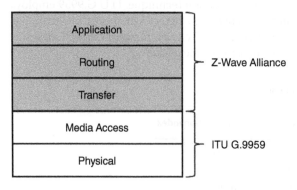

Figure 2.7 Z-Wave Protocol Stack

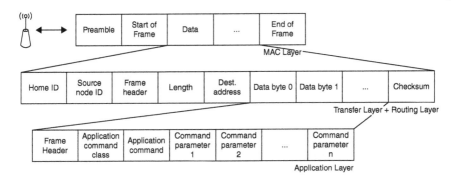

Figure 2.8 Z-Wave Packet Layers [16]

standardized by an independent standards body follows the trend of other wireless IoT protocols. The Z-Wave Alliance provides specifications for the operations of the upper layers.

The lower two layers of Z-Wave were standardized by the International Telecommunication Union as ITU G.9959 in 2012 [9]. That the lower layers are standardized by an independent standards body follows the trend of other wireless IoT protocols. The Z-Wave Alliance provides specifications for operations of the upper layers.

2.4 IEEE 802.15.4

IEEE 802.15.4 is the IEEE standard covering Low Rate Wireless Personal Area Networks (LR-WPAN). It is important to note that IEEE 802.15.4 itself is not a wireless IoT implementation or technology. It is a standard to which technologies conform. Among the wireless IoT technologies based on IEEE 802.15.4 are the ZigBee and Thread specifications. The protocol stack for IEEE 802.15.4 is shown in Figure 2.9 [17]. The Logical Link Control (LLC) of the 802.15.4 stack is defined by IEEE 802.2. This is similar to Bluetooth (802.15.1) where the stack allows for an IEEE 802.2 defined LLC.

IEEE 802.15.4 was first standardized into IEEE 802.15.4-2003 [18]. This first version specified the channel plan and the Binary Phase-Shift Keying (BPSK), Offset Quadrature Phase-Shift Keying (OQPSK) modulation schemes. Parallel-Sequence Spread Spectrum was added to the standard when the standard was updated to IEEE 802.15.4-2006 [19]. The ZigBee specification has an option to make use of that modulation scheme. The standard has since been updated to IEEE 802.15.4-2011 [20] and then to IEEE 802.15.4-2015 [21]. New waveforms and features have been added.

There is a singular frequency channel plan for all protocols based on IEEE 802.15.4. IEEE 802.15.4 identifies three frequency bands. Those are the 2.4 GHz

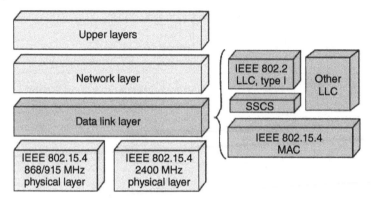

Figure 2.9 IEEE 802.15.4 Protocol Stack [17]

international Industrial, Medical, and Scientific (ISM) band, the 915 MHz ISM band, and the 868 MHz Short Range Device band (SRD). The sub-1 GHz bands are not international. These sub-1 GHz unlicensed bands are defined by individual regions. The SRD band is available in Europe. The sub-1 GHz band is available in the United States.

IEEE 802.15.4 standard specifies 27 channels, numbered 0 to 26, across these three frequency bands. The frequencies for those channels are given in Table 2.9. The channel plan is illustrated in Figure 2.10 [17].

Reference [17] shows the layout of the IEEE 802.15.4 packet across the media access and physical layer as defined by the standard. The media access (MAC) layer controls message sequencing, error detection, and addressing. The physical (PHY) layer includes synchronization, modulation, and spreading.

2.4.1 Transceiver

The receiver sensitivity required by IEEE 802.15.4 varies with the modulation type and band of operation. At a specified receiver sensitivity, the receiver is expected to operate with a 1% Packet Error Rate (PER) or less. The sensitivities are listed in Table 2.10. The channel plan is illustrated in Figure 2.10.

Table 2.9 IEEE 802.15.4 Channel Plan

Channels	Range
0	868.3 MHz
1–10	$904 + 2^*n$ MHz
11–26	$2350 + 5^*n$ MHz

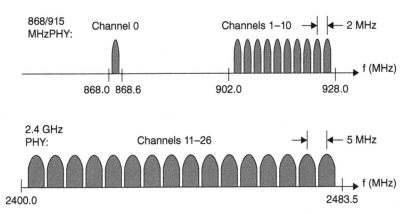

Figure 2.10 802.15.4 Channel Plan [17]

IEEE 802.15.4 recommends a transmit power of 0 dBm.

These values can be used to calculate a link budget and link margin. Link budgets and margin are discussed in Chapter 3.

2.4.2 Frequency Channels

IEEE 802.15.4 can operate in the 2.4 GHz Industrial, Scientific, and Medical (ISM) band or the sub-1 GHz unlicensed bands that are defined by different regions. The operation in the 2.4 GHz band allows IEEE 802.15.4 vendors to produce solutions that are viable internationally. The 2.4 GHz ISM band is very congested and this congestion must be taken into account. The fact that IEEE 802.15.4 can select one of many channels for operations in the 2.4 GHz ISM band allows it to choose an uncontested, or at least less contested, frequency channel. IEEE 802.15.4 uses Direct-Sequence Spread Spectrum to mitigate in-channel interference.

Table 2.10 IEEE 802.15.4 Receiver Sensitivities

Band	Modulation	Sensitivity
2.4 GHz	OQPSK	−85 dBm
Sub-1 GHz	BPSK	−92 dBm
Sub-1 GHz	ASK	−85 dBm
Sub-1 GHz	OQPSK	−85 dBm

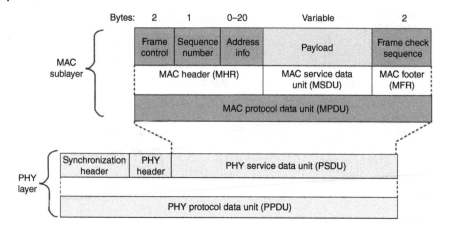

Figure 2.11 IEEE 802.15.4 Packet Structure [17]

Operating in the sub-1 GHz band allows IEEE 802.15.4 better indoors propagation and allows IEEE 802.15.4 to operate in a less congested environment. The channel plan is illustrated in Figure 2.10.

2.4.3 Typical Range

The operational range of a wireless system depends on many factors. Chapter 3 will review link budgets, which are necessary to calculate operational range in a given environment. With that caveat in place, the typical operating range can be discussed. Reference [12] lists typical operating ranges for various wireless IoT protocols. Reference [12] specifically refers to the ZigBee specification and not IEEE 802.15.4; however, it is IEEE 802.15.4 that dictates the transceiver. Therefore, based on [12], IEEE 802.15.4 offers typical ranges between 10 meters and 100 meters.

2.4.4 Access and Spread Spectrum

IEEE 802.15.4 uses Frequency Division Multiple Access (FDMA) for channel definitions. This singular frequency channel is shared by all nodes in a given IEEE 802.15.4 network. FDMA is discussed in Chapter 5.

IEEE 802.15.4 uses two spread spectrum techniques, depending on the band and modulation for that particular IEEE 802.15.4 network. Those two techniques are Direct-Sequence Spread Spectrum (DSSS) and Parallel-Sequence Spread Spectrum (PSSS). The DSSS spreading is intended to mitigate interference. Both of these concepts are discussed in Chapter 5. The use of the different spread spectrum techniques is listed in Table 2.11.

Table 2.11 IEEE 802.15.4 Spread Spectrum Techniques

Band	Modulation	Spreading
2.4 GHz/915 MHz/868 MHz	OQPSK	DSSS (16-ary)
868 MHz/915 MHz	ASK	PSSS
868 MHz/915 MHz	BPSK	DSSS (binary)

The DSSS technique used for the IEEE 802.15.4 OQPSK modulation scheme is sometimes called "16-ary orthogonal." Specifically, the IEEE 802.15.4 OQPSK modulation scheme takes four input bits, maps those to 16 "chip" bits, and then processes those chips through the OQPSK modulator. This concept is discussed in Chapter 4.

Nodes in a given IEEE 802.15.4 network use the same spreading code. To avoid collision within a network, IEEE 802.15.4 uses either time-slots (Time-Division Multiple Access) or Carrier-Sense Multiple Access. Both of these concepts are discussed in Chapter 5.

2.4.5 Modulation and Data Rate

IEEE 802.15.4 utilizes different physical layers depending on the band of operation. Those values are shown in Table 2.12.

OQPSK stands for Offset Quadrature Phase-Shift Keying. OQPSK is discussed in Chapter 4. The IEEE 802.15.4 standard uses a sinusoidal pulse shaping in the specified OQPSK modulation scheme. That pulse shaping makes the OQPSK modulation much like Minimum-Shift Keying (MSK). MSK and pulse shaping are discussed in Chapter 4.

ASK stands for Amplitude-Shift Keying. ASK is discussed in Chapter 4. IEEE 802.15.4 defines an m-ary ASK modulation combined with a Parallel-Sequence Spread Spectrum technique. This modulation scheme is only used in sub-1 GHz channels. This scheme is used in the ZigBee specification.

Table 2.12 IEEE 802.15.4 Modulations and Data Rates

Band	Modulation	Data rate
2.4 GHz/915 MHz	OQPSK	250 kb/s
868 MHz	OQPSK	100 kb/s
868 MHz/915 MHz	ASK	250 kb/s
868 MHz	BPSK	20 kb/s
915 MHz	BPSK	40 kb/s

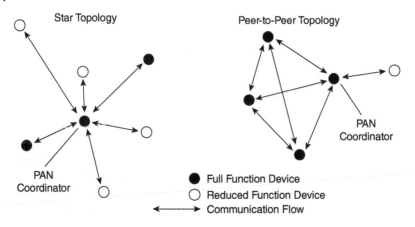

Figure 2.12 IEEE 802.15.4 Star and Peer-to-Peer Topologies [21]

2.4.6 Error Detection and Correction

IEEE 802.15.4 uses no forward error correction technique. IEEE 802.15.4 defines cyclic redundancy checks to detect whether a bit error has occurred in a received packet. Because there is no forward correction, a single bit error will require retransmission of the packet.

Cyclic redundancy checks and automatic repeat requests are discussed in Chapter 5.

2.4.7 Network Topology

IEEE 802.15.4 identifies two topologies: star and "Peer-to-Peer." These are illustrated in Figure 2.12. The "Peer-to-Peer" topology is a mesh topology.

IEEE 802.15.4 also lists a "cluster tree" as a use case of the mesh topology. The cluster tree is illustrated in Figure 2.13 from [21]. The more complex tree topology is formed by having multiple coordinators in a mesh network.

2.5 The ZigBee Specification

The ZigBee specification was originally developed by the Ember Corporation to allow a wireless IoT network to be populated with a large number of nodes. The result of this development was a low-power low-data-rate WPAN that could incorporate more nodes than Bluetooth by an order of magnitude [22]. IEEE 802.15.4 was ratified in 2003. ZigBee Specification 1.0 was made available by the ZigBee Alliance in 2005. Silicon Labs acquired Ember in 2012.

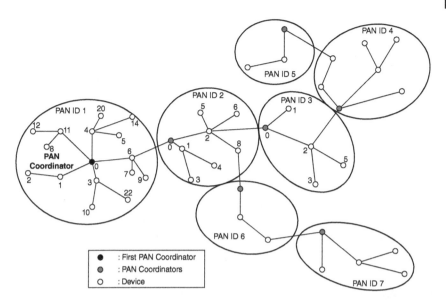

Figure 2.13 IEEE 802.15.4 Cluster Tree Topology [21]

The ZigBee specification is tied closely to the IEEE 802.15.4 standard, but the two are not synonymous [23]. The IEEE 802.15.4 standard offers a wide range of physical layer specifications, only a subset of which is standardized under the ZigBee specification. The protocol stack for the ZigBee specification is illustrated in Figure 2.14, provided by reference [12]. Figure 2.14 shows that the

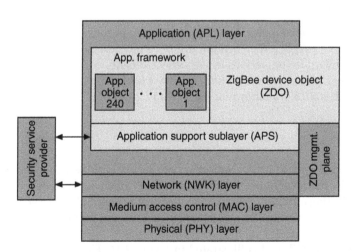

Figure 2.14 ZigBee Specification Protocol Stack [12]

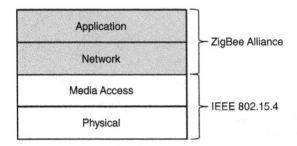

Figure 2.15 Simplified ZigBee Specification Protocol Stack

ZigBee specification provides security functionality across layers. A simplified version of the ZigBee specification protocol stack is shown in Figure 2.15. The IEEE 802.15.4 standard defines the physical and media access layers [18]. Much like other IoT protocols, only the lower layers are maintained by an independent standards body.

- The physical layer handles frequency conversion, modulation, and spreading.
- The media access layer handles Carrier Sense Multiple Access (CSMA), Time-Division Multiple Access (TDMA) in the form of slotting, error detection, and retransmission.
- The network layer handles network discovery, network management, and routing.
- The application layer is the seat of the application. The ZigBee specification provides a uniform support sublayer such that numerous higher layer standards can be built upon the wireless links and networking can be established in the lower layers.

The ZigBee specification operates in multiple bands, as specified by the standard IEEE 802.1.4. The ZigBee specification utilizes a physical layer from the IEEE 802.105.4 standard for operations in each selected band.

The layout of the ZigBee packet across the media access and physical layers follows that of IEEE 802.15.4, as shown in Figure 2.11.

2.6 Thread

Thread was developed by Nest Labs, a subsidiary of Alphabet Inc. (which owns Google). Nest Labs had experience in the development of smart devices for smart home applications. In 2014, Nest Labs joined with other industry partners to form the Thread Group [24], which is an alliance that maintains and promotes the Thread protocol.

The protocol stack for Thread is illustrated in Figure 2.16. Figure 2.16 has been recreated from [25]. As can be seen in Figure 2.16, Thread sits atop the

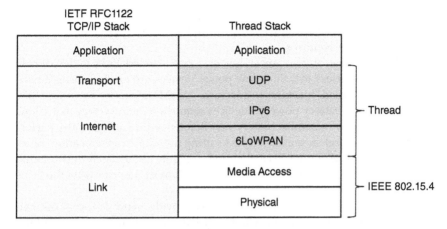

Figure 2.16 Thread Protocol Stack Compared to TCP/IP Stack

IEEE 802.15.4 standards as does the ZigBee specification. Much like the Zig-Bee specification, the Thread Group maintains the upper layers of the protocol while IEEE maintains the lower layers. The physical and MAC layers for Thread are defined by IEEE 802.15.4 OQPSK PHY operating in the 2.4 GHz ISM band.

The Thread protocol stack follows the model of the TCP/IP reference discussed in Chapter 1. The duties of the transport layer are fulfilled by the "User Datagram Protocol" (UDP) [26]. UDP is a connectionless transport protocol that fulfills the duties of the transport layer. The duties of the internet layer are fulfilled using Internet Protocol version 6 (IPv6) [27] and "IPv6 over Low-Power Wireless Personal Area Networks" (6LoWPAN) [28].

2.7 Wi-Fi

The term "Wi-Fi" has become ubiquitous enough to be synonymous with "wireless internet." But what exactly is "Wi-Fi"? "Wi-Fi" stands for "wireless fidelity" and it is not a wireless link but rather a trademark of the Wi-Fi Alliance. The Wi-Fi Alliance is a non-profit organization that certifies wireless products as Wi-Fi interoperable. The actual wireless protocol used for that Wireless Local Area Network (WLAN) is defined by the IEEE 802.11 specification [29]. The IEEE 802.11 standards family provides standards for the PHY and MAC layers of Wireless Local Area Networks. Much like reference [29], this book has treated the two terms as synonymous for the convenience of the reader. The book will continue to treat the two terms as synonymous for the convenience of the readers, e.g., when discussing a Wi-Fi WLAN. The IEEE 802.11 standards were first published in 1997 [30]. The IEEE 802.11 standards have been amended many times over the years. The Wi-Fi Alliance only certifies a subset of the waveforms contained in the standards. That is to say that although the IEEE 802.11 defines several waveforms, the link we use as "Wi-Fi" only employs a subset of those waveforms.

This book focuses on low power, low data rate, wireless links in device networks for the wireless IoT. Such low power devices are used to form "device networks," also called "machine-to-machine" (M2M) networks. Forming M2M networks among battery powered devices requires a radio system that is low energy in order to maximize battery life. Reference [31] analyzes the performance of three wireless sensor networks using the ZigBee specification, Bluetooth, and Wi-Fi. While Wi-Fi can be used, the issue immediately encountered is energy consumption. Therefore, Wi-Fi does not meet the criteria for the focus of this book.

Nevertheless, the IEEE 802.11 standards provide many IoT services and commonly provide gateways for wireless IoT device networks. Because of the importance of the IEEE 802.11 standards to the wireless IoT, it is necessary to provide some background on those standards.

The stack for Wi-Fi, as a means to establish a WLAN, is shown in Figure 2.17. The stack follows the TCP/IP stack, as discussed in Chapter 1, and as defined by reference [32] (IETF RFC1122). The lowest layer of the TCP/IP stack, the Link Layer, is defined by IEEE standards. The upper layers are defined by the IETF. The lowest layer is composed of the physical layer, the MAC layer, and the Data

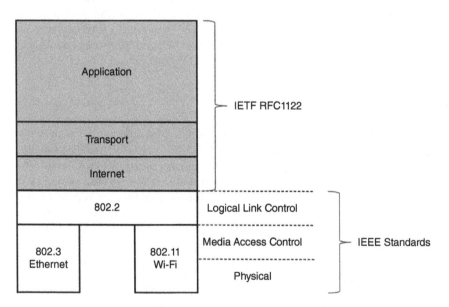

Figure 2.17 IEEE 802.11 WLAN Stack

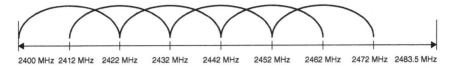

2400 MHz 2412 MHz 2422 MHz 2432 MHz 2442 MHz 2452 MHz 2462 MHz 2472 MHz 2483.5 MHz

Figure 2.18 Overlapping Wi-Fi Channels [33]

Link layer. In Figure 2.17, IEEE 802.11 (WLAN) and IEEE 802.3 (Ethernet) are shown side by side. Both sets of standards define the physical and MAC layers. Both sets of standards interface with the Data Link layer, as defined by the IEEE 802.2 standard. Both IEEE 802.3 and IEEE 802.11 can be used to create a LAN. The role of IEEE 802.11 is to provide a wireless replacement for physical cables.

Wi-Fi operates in the 2.4 and 5 GHz ISM bands. Either band of operation can provide a gateway for a device network. Wi-Fi operation in the 2.4 GHz band is a source of interference for many device networks, and that will be discussed in Chapter 5.

Wi-Fi channels in the 2.4 GHz band begin with channel 1 at 2412 MHz and continue to channel 13 at 2472 MHz with 5 MHz between each channel. The complicated part of this channel plan is that the channels are much wider than 5 MHz. This means that the channels overlap, as shown in Figure 2.18 [33]. A set of non-overlapping channels are shown in Figure 2.19 [33]. There is an additional channel 14 centered at 2484 MHz, but the availability of that last channel on the edge of the 2.4 GHz ISM band depends upon the specific region of operation. Because the Wi-Fi channels overlap, non-overlapping sets of channels are chosen to be used. Channels 1, 6, and 11 are commonly chosen in the United States.

Wi-Fi employs two physical layers, Orthogonal-Frequency Division Multiplexing (OFDM) and Direct-Sequence Spread Spectrum. Because Wi-Fi is not within the focus of this book, and no other wireless IoT protocol uses OFDM, the theoretical concepts behind OFDM will be left to other texts. The prevalence of Wi-Fi OFDM does affect radio manufacturing and this is briefly discussed in Chapter 3. DSSS is used by wireless IoT protocols such as the ZigBee specification. DSSS will be discussed in Chapter 4.

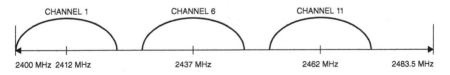

CHANNEL 1   CHANNEL 6   CHANNEL 11

2400 MHz 2412 MHz   2437 MHz   2462 MHz   2483.5 MHz

Figure 2.19 Non-Overlapping Wi-Fi Channels [33]

References

1 A. S. Tanenbaum, *Computer Networks*. Upper Saddle River: Prentice Hall, 2003.

2 Part 15.1: Wireless Medium Access Control (MAC) and Physical Layer (PHY) Specifications for Wireless Personal Area Networks (WPANs), IEEE 802.15.1-2002, 2002.

3 Part 15.1: Wireless Medium Access Control (MAC) and Physical Layer (PHY) Specifications for Wireless Personal Area Networks (WPANs), IEEE 802.15.1-2005, 2005.

4 Bluetooth Core Specification version 5.0, Bluetooth Special Interest Group, 2016.

5 S. Raza, P. Misra, Z. He, and T. Voigt, "Bluetooth smart: An enabling technology for the Internet of Things," in *2015 IEEE 11th International Conference on Wireless and Mobile Computing, Networking and Communications (WiMob)*, Abu Dhabi, United Arab Emirates, Oct. 2015, pp. 155–162.

6 J. Padgette, K. Scarfone, and L. Chen, "Guide to Bluetooth Security: Recommendations of the National Institute of Standards and Technology," *NIST Special Publication*, Vols. 800-121, 2012.

7 Mesh Model: Bluetooth® Specification, Bluetooth Special Interest Group, Mesh Working Group, 2017.

8 K.-H. Chang, "Bluetooth: A viable solution for IoT?," *IEEE Wireless Commun.*, vol. 21, no. 6, pp. 6–7, 2016.

9 Short range narrow-band digital radio communication transceivers—PHY and MAC layer specifications, Recommendation ITU-T G.9959, 2012.

10 Short range narrow-band digital radio communication transceivers—PHY, MAC, SAR and LLC layer specifications, Recommendation ITU-T G.9959, 2015.

11 A. Brandt and J. Buron, RFC 7428: Transmission of IPv6 Packets over ITU-T G.9959 Networks, Internet Engineering Task Force, 2015.

12 C. Gomez and J. Paradells, "Wireless home automation networks: A survey of architectures and technologies," *IEEE Commun. Mag.*, vol. 48, no. 6, pp. 92–101, 2010.

13 A. Westervelt. (2012, Mar. 21). Could smart homes keep people healthy? *Forbes* [Online]. Available: https://www.forbes.com/sites/amywestervelt/2012/03/21/could-smart-homes-keep-people-healthy/#2593a254579a

14 A. Stafford. (2005, Dec. 29). First look: Catch the home automation Z-Wave. *PC World*. Available: https://www.techhive.com/article/123856/article.html

15 M. B. Yassein, W. Mardini, and A. Khalil, "Smart homes automation using Z-Wave protocol," in *2016 IEEE Int. Conf. Eng. MIS (ICEMIS)*, Agadir, Morocco, Sep. 2016, pp. 1–6.

16 P. Amaro, R. Cortesão, J. Landeck, and P. Santos, "Implementing an advanced meter reading infrastructure using a Z-Wave compliant wireless sensor network," in *IEEE Proc. 2011 3rd Int. Youth Conf. Energetics*, Leiria, Portugal, Jul. 2011, pp. 1–6.

17 E. Callaway, P. Gorday, L. Hester, J. A. Gutierrez, M. Naeve, B. Heile, and V. Bahl, "Home networking with IEEE 802.15.4: A Developing standard for low-rate wireless personal area networks," *IEEE Commun. Mag.*, vol. 40, no. 8, pp. 70–77, 2002.

18 Part 15.4: Wireless Medium Access Control (MAC) and Physical layer (PHY) specifications for Low-Rate Wireless Personal Area Networks (LR-WPANs), IEEE 802.15.4-2003, 2003.

19 Part 15.4: Wireless Medium Access Control (MAC) and Physical layer (PHY) specifications for Low-Rate Wireless Personal Area Networks (LR-WPANs), IEEE 802.15.4-2006, 2006.

20 Part 15.4: Low-Rate Wireless Personal Area Networks (LR-WPANs), IEEE 802.15.4-2011, 2011.

21 Part 15.4: Low-Rate Wiress Networks, IEEE 802.15.4-2015, 2015.

22 E. Corcoran. (2014, Sep. 6). Giving voice to a billion things. *Forbes* [Online]. Available: https://www.forbes.com/free_forbes/2004/0906/144d.html

23 A. Kumar, A. Sharma, and K. Grewal, "Resolving the paradox between IEEE 802.15. 4 and Zigbee," in *IEEE 2014 Int. Conf. Optimization Rel. Inform. Technol. (ICROIT)*, Faridabad, India, Feb. 2014, pp. 484–486.

24 N. Randewich, *Google's Nest launches network technology for connected home*, Reuters, July 15, 2014 .

25 S. A. Al-Qaseemi, M. F. Almulhim, H. A. Almulhim, and S. R. Chaudhry, "IoT architecture challenges and issues: Lack of standardization," in *IEEE Future Technol. Conf.*, San Francisco, CA, Dec. 2016, pp. 731–738.

26 J. Postel, RFC 768: User Datagram Protocol, The Internet Engineering Task Force, 1980.

27 S. Deering and R. Hinden, RFC 2460: Internet Protocol, Version 6 (IPv6), The Internet Engineering Task Force, 1998.

28 G. Montenegro, N. Kushalnagar, J. Hui and D. Culler, RFC 4944: Transmission of IPv6 Packets over IEEE 802.15.4 Networks, The Internet Engineering Task Force, 2007.

29 E. Ferro and F. Potorti, "Bluetooth and Wi-Fi wireless protocols: A survey and a comparison," *IEEE Wireless Commun.*, vol. 12, no. 1, pp. 12–26, 2005.

30 Part 11: Wireless LAN Medium Access Control (MAC) and Physical layer (PHY) specifications, IEEE 802.11-1997, 1997.

31 G. Mois, S. Folea, and T. Sanislav, "Analysis of three IoT-based wireless sensors for environmental monitoring," *IEEE Trans. Instrum. Meas.*, vol. 66, no. 8, pp. 2056–2064, 2017.

32 R. Braden, "RFC1122: Requirements for Internet hosts – communication layers," The Internet Engineering Task Force, 1989.

33 Part 11: Wireless LAN Medium Access Control (MAC) and Physical layer (PHY) specifications: Higher-speed physical layer extension in the 2.4 GHz band, IEEE 802.11b-1999, 1999.

3

Radio Layer

The Internet of Things (IoT) covers a wide array of applications. The protocols defined by the wireless standards employed by a given IoT application are chosen to facilitate that application. This concerted effort to empower IoT applications begins at the very bottom of the protocol stack.

The physical layer, also called layer 1 or simply the "PHY", is the lowest layer in the stack of layers of a given protocol. It is in the PHY that digital data is given a physical form, which can be transmitted, propagated, and received. The exact boundary of the physical layer varies amongst protocol standards. In general, the physical layer is concerned with propagating data across a medium, which in the case of the wireless IoT is the wireless spectrum. Therefore, physical layer specifications standardize things like receiver sensitivity, link budgets, channel models, waveforms, and error rates.

Chapter 1 presented a unified model of a stack of the lower layers of the wireless IoT. That stack is shown again in Figure 3.1. As this book progresses, that stack will be traversed from the bottom up. The placement of this chapter is illustrated by a lack of shading and an arrow.

This book has broken the physical layer into two components. It is beyond the scope of one chapter, or even one book, to cover the entire breadth and depth of physical layer considerations for wireless communications. This chapter will explain the concepts relevant to the physical layers of the standards related to the Internet of Things, present the various physical layer choices available, explain what is advantageous about those choices, and provide a list of useful references for further study.

3.1 The Wireless System

Wireless systems are often described in complex terms that are specific to one standard or another. It is important to start this chapter with a simplified model of a wireless system that lends itself to all other models. To that end, a wireless system can be defined as being comprised of three basic components: A

The Wireless Internet of Things: A Guide to the Lower Layers, First Edition. Daniel Chew.
© 2019 by The Institute of Electrical and Electronic Engineers, Inc. Published 2019 by John Wiley & Sons, Inc.

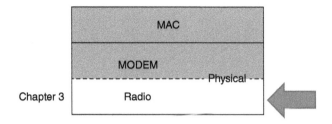

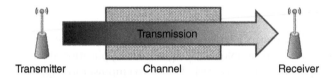

Figure 3.1 Traversing the Stack: The Radio Layer

Figure 3.2 Communication System Diagram

transmitter (Tx), a channel, and a receiver (Rx). The relationship between these components is shown in Figure 3.2. The transmitter emits a modulated signal. That modulated signal propagates through the channel. The channel imparts effects upon the signal, such as the loss of power over distance. The receiver then attempts to recover the signal. A transceiver is a device that contains both a transmitter and a receiver.

A wireless standard will specify rules for the transmitter and models for the channel. Receiver details are largely left to each competing development company. Remember that the standard was written with the intention of receivers recovering the transmitted signal. No matter how dense the standard may seem, the signal was meant to be demodulated.

Wireless standards for the Internet of Things follow this same logic. Wireless standards are chosen for an IoT application because that standard enables technology that facilitates the application. Examples include low power transmissions, ease of channel estimation, and low cost receivers.

3.2 Basic Transceiver Model

In order to discuss the reasoning behind the wireless standards of the IoT, some basic model of the technology that empowers those standards needs to be established. The basic model of modern transceivers has three components: Analog RF front-end, digital channelization, and baseband controller. Figure 3.3 illustrates a basic model of modern transceivers.

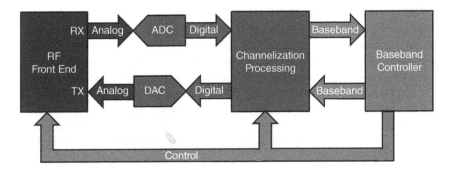

Figure 3.3 Transceiver Functional Block Diagram

A received signal flows from the analog front-end to an analog-to-digital converter (ADC). This received signal may contain several potential signals of interest. Therefore, the signal at this stage will be referred to as a "bandwidth of interest." The analog front-end must provide initial selection from a band of operation to a smaller bandwidth of interest. The ADC performs quantization and sampling. These concepts are covered in a number of texts. The explanation in reference [1] is particularly approachable.

After the ADC has performed quantization and sampling, digital signal processing begins. The initial digital signal processing is done at a high sample rate. This is where digital channelization occurs. After the bandwidth of interest has been reduced to a singular signal of interest, lower sample rate "baseband" processing begins. For the transmit chain, the process is reversed. A signal is generated at baseband and then channelized. The channelized digital signal is then converted to an analog signal through a digital-to-analog converter (DAC). The following sections will provide details and nuance to these components.

3.2.1 The Analog Front-End

The analog front-end is the RF circuit designed to perform analog signal conditioning of an RF signal before digitization. The analog front-end is the connection between the antenna and digitization (ADC/DAC).

The role of the analog front-end is to condition a received signal for digitization and condition a generated signal for transmission. The problem encountered is that the bandwidth of interest (BOI) may not be within the Nyquist bandwidth of the digital converters. The Nyquist bandwidth is defined by the sampling rate of the front-end. Terms like "Nyquist bandwidth" and "Nyquist frequency" are products of the Nyquist Sampling Theorem. References [1, 2], and [3] can provide more information on the Nyquist Sampling Theorem, but it will be addressed here briefly. The Nyquist Sampling Theorem defines the minimum sample rate required in order to avoid "aliasing" when sampling a bandwidth. Aliasing is a type of distortion that occurs when frequency components in a bandwidth to be sampled exceed the Nyquist frequency. The Nyquist frequency is one-half the sampling rate. The Nyquist bandwidth is the bandwidth that can be sampled without aliasing.

There are three common solutions to this problem. Those are superheterodyne, direct conversion, and RF sampling. These three RF front-end architectures will be explored in the following sections. A basic block diagram will be provided for each, and common pitfalls will be discussed. Direct-conversion architectures are of particular interest to IoT and will be explored in greater detail.

Unfortunately, it is outside the scope of this book to delve deeply into the design of RF front-ends. There are numerous texts that engage in this task. If the reader is interested in learning how to design a functioning RF front-end, then excellent texts can be found in [4, 5], and [6]. There are numerous design considerations for RF circuits. True appreciation of these design considerations spans multiple books and years of study. The scope of this section is to introduce the reader to common front-end architectures and relate those architectures to IoT standards.

3.2.1.1 Superheterodyne

The superheterodyne (superhet) receiver was first proposed in 1921 [7]. Figure 3.4 shows a simplified block diagram of a superhet receiver. The signal is processed at two frequency stages called the "Radio Frequency" (RF) and the "Intermediate Frequency" (IF). RF is regarded as "high" and requires expensive circuitry for analog signal conditioning. The transceiver performs most of the signal conditioning at IF. This allows the majority of the analog circuitry to perform analog signal conditioning at lower frequencies. By way of varying the frequency of the local oscillator the user can select a narrow bandwidth of interest from a wider initially received band. The Local Oscillator (LO) provides a local sinusoid to select the bandwidth of interest. A pre-selection

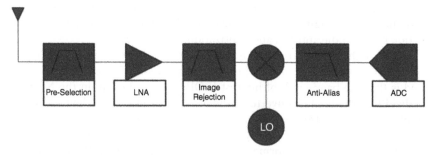

Figure 3.4 Superheterodyne Receiver

filter prevents out-of-band interference. A Low Noise Amplifier (LNA) provides gain at the front of the receiver. LNAs are amplifiers with a low noise figure. Noise figure is a measure of how much the signal-to-noise ratio degrades through the device. More information on noise figures can be found in references [4] and [2]. A bandpass image rejection filter is placed before the mixing stage to prevent image frequencies from corrupting the down-converted output. The image rejection filter will be discussed later in this section. Finally, an anti-alias filter serves a dual purpose. The low-pass anti-alias filter prevents aliasing at the Analog-to-Digital Converter (ADC) and provides a channel selection filter. A channel selection filter is necessary in a superhet because of the real mixing.

The mixing stage for the superhet is real valued. Equation (3.1) shows the trigonometric identity used for real-valued mixing.

$$\cos \phi \cos \theta = \frac{1}{2} \cos (\phi - \theta) + \frac{1}{2} \cos (\phi + \theta) \tag{3.1}$$

The result pushes the signal both up and down in frequency. Because the desired signal is pushed both up and down in frequency, the channel selection filter is necessary to remove the unwanted product. The desired product, which is the down-converted product in the case of the receiver in Figure 3.4, is allowed to pass through the low-pass filter.

Analog mixers, emulating multiplication, are not perfect. Some of the energy of the LO will be leaked into the output at the LO frequency. Therefore, LO frequencies must be chosen outside the bandwidth of interest at the IF frequency.

The image rejection filter is necessary because of other signals in the bandwidth of interest. The real-valued mixing process also mixes other signals up and down such that there is an undesired "image frequency" that will move into the same position as the desired signal. This process is shown in Figure 3.5. In order to understand what is happening with image frequencies, one needs to examine the two-sided spectrum. Figure 3.5 shows the two-sided spectrum of a desired signal and an undesired signal. The heterodyne real-valued mixing process will push a copy of both signals high and low in frequency. The desired and undesired are then situated such that the negative frequency copy of the undesired signal collides with the desired signal, which is being down-converted.

Figure 3.6 shows equations for the example illustrated in Figure 3.5. The desired signal is situation at frequency "10." The LO is tuned to frequency "9." It is desired to down-convert the desired signal to frequency "1." There is an undesired signal at frequency "8." The mixing process moves copies of the desired signal to frequencies "19" and "1." The channel selection filter will filter out the up-converted product. The undesired signal will mix to frequencies "17" and "1." Therefore, the undesired signal must be filtered out before the mixing stage.

The digitization process also requires special considerations. Figure 3.7 shows the two-sided spectrum of a bandwidth to be sampled. That bandwidth to be sampled contains two signals, the spectrums of which are represented

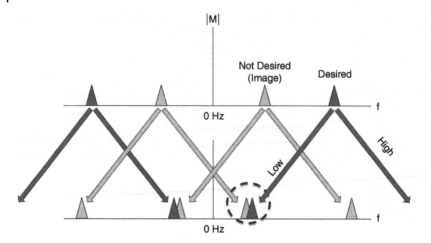

Figure 3.5 Real-Mixing Images Frequencies

Signal   LO   High   Low

$$\cos(2\pi 10t) * \cos(2\pi 9t) = \frac{1}{2}\cos(2\pi 19t) + \frac{1}{2}\cos(2\pi t)$$

$$\underbrace{\cos(2\pi 8t) * \cos(2\pi 9t)}_{\text{Images}} = \frac{1}{2}\cos(2\pi 17t) + \underbrace{\frac{1}{2}\cos(2\pi t)}_{\text{Images}}$$

Figure 3.6 Image Frequency Example

Figure 3.7 Real-Valued ADC Bandwidth

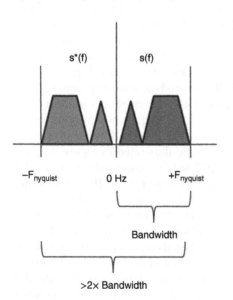

by a triangle and a trapezoid. There are three vertical lines denoting important values in the frequency domain in Figure 3.7. Those are 0 Hz, + $F_{nyquist}$, and $-F_{nyquist}$. $F_{nyquist}$ is the "Nyquist frequency," which is one-half of the sampling rate.

The Nyquist frequency demarcates the highest value for a frequency component in the bandwidth to be sampled without "aliasing." It is often stated that the Nyquist Sampling Theorem requires the minimum sampling rate to be twice the bandwidth to be sampled; however, this applies to a one-sided bandwidth. The reason that there is both a + $F_{nyquist}$ and a $-F_{nyquist}$ is because this figure shows the two-sided spectrum. A two-sided bandwidth contains an equal range of positive and negative frequencies. For real-valued signals, the two-sided bandwidth is twice the one-sided bandwidth. For a two-sided bandwidth, the minimum required sampling rate to avoid aliasing is equal to the two-sided bandwidth. The two-sided spectrum of a real-valued signal consists of the spectrum of that signal in positive frequencies and a complex-conjugate version of the same spectrum in negative frequencies. This is a symmetry that occurs in the spectrum of real-valued signals. This type of symmetry is called "conjugate" or "Hermitian" symmetry.

As shown in Figure 3.8, the bandwidth of interest is down-converted to baseband. The bandwidth being digitized must be equal to or less than one-half of the sampling rate. If the bandwidth of interest were placed perfectly so as to be

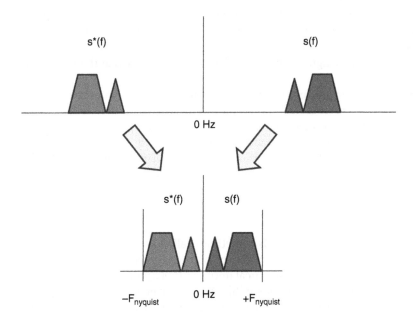

Figure 3.8 Perfect Real-Valued Down-Conversion

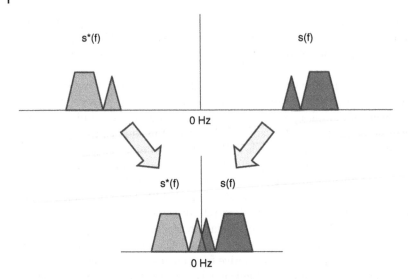

Figure 3.9 Overlapping Bandwidths by Real-Valued Down-Conversion

centered at 0 Hz, then the relationship illustrated in Figure 3.8 could be used. The two sides of the spectrum in Figure 3.8 show conjugate symmetry.

Unfortunately, down-conversion is never so precise. The bandwidth of interest will miss the mark. Carrier synchronization has not yet been performed. This will result in some aliasing or overlapping with the conjugate, as viewed in the two-sided spectrum shown in Figure 3.9. This also shows that for real-valued signals, the two-sided bandwidth is twice the one-sided bandwidth.

The solution is to move the bandwidth of interest to an "intermediate frequency" (IF), from where additional baseband processing may follow. With the use of such an intermediate frequency, the bandwidth of interest is placed away from 0 Hz inside the total bandwidth to be sampled. This is shown in Figure 3.10. As shown in Figure 3.10, sampling the signal at this intermediate frequency will necessitate a higher sampling rate. This makes the superhet less bandwidth efficient and more expensive.

3.2.1.2   Direct Conversion

A direct-conversion transceiver is sometimes called homodyne or Zero-IF or a quadrature transceiver. The concept for homodyne receiver has existed for many years, but has not been practical until the 1990s with the advent of more advanced radio receiver technology [8]. Direct-conversion has been successfully implemented in Radio Frequency Integrated Circuits (RFIC). Extensive details on direct-conversion analog front-ends can be found in [6] and [5].

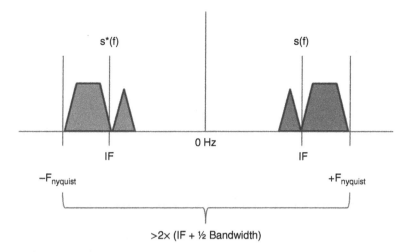

Figure 3.10 Sampling at IF

Figure 3.11 shows a simplified block diagram of a Direct-Conversion Receiver (DCR). The bandwidth of interest is processed in only one frequency stage. This single stage mixes the bandwidth of interest to baseband. By way of varying the frequency of the local oscillator the user can select a narrow bandwidth of interest from a wider initially received band. The Local Oscillator (LO) provides two local sinusoids, equal in frequency but separated by 90 degrees, to select the bandwidth of interest. A pre-selection filter prevents out-of-band interference. A Low Noise Amplifier (LNA) provides gain at the front of the receiver. The phase information of the bandwidth of interest is not known a priori. Therefore, a DCR has two arms, the in-phase (real) and quadrature (complex) phase versions of the down-converted signal. This two-arm approach to mixing is referred to as "quadrature mixing." The digitization must, therefore, use two

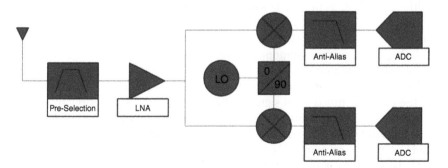

Figure 3.11 Direct-Conversion Receiver

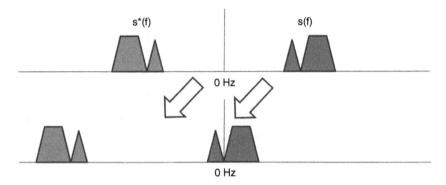

Figure 3.12 Complex-Valued Mixing

channels. This technique is referred to as complex-valued or quadrature sampling, as opposed to "real-valued" sampling in a superhet.

This architecture takes advantage of Euler's formula (equation 3.2) to create a complex-valued oscillator and mixing stage. This concept of complex-valued mixing is illustrated in Figure 3.12. The mixing process moves the bandwidth of interest in only one direction, either up or down. Figure 3.12 shows a real-valued signal becoming a complex-valued signal by way of this process. The two-sided spectrum of the initial real-valued signal shows conjugate symmetry. The complex-valued down-converted version in Figure 3.12 does not have conjugate symmetry.

$$e^{j\theta} = \cos(\theta) + j\sin(\theta) \tag{3.2}$$

One benefit of this method is that the sampled bandwidth is doubled since positive and negative frequencies now carry unique information. This is illustrated in Figure 3.13. Compare Figure 3.13 to Figure 3.10. Note that the bandwidth of the ADC is more efficiently used. The bandwidth of the signal sampled may now be equal to the Nyquist frequency, assuming the signal could be down-converted perfectly to 0 Hz. This does not violate the sampling theorem, as the largest frequency component is still bound by the Nyquist frequency. The signal now straddles the y-axis of the spectrum. While the sampling rate may be lower, the ADCs must now generate twice the data. Therefore, the data rate for the perfectly tuned DCR signal is the same as that for the perfectly tuned superhet. Also note that because there is no conjugate component overlapping with the bandwidth of interest around 0 Hz, the bandwidth of interest can be safely set slightly off from 0 Hz.

Another benefit is the reduction of analog components. The image rejection and channel selection filters from the superheterodyne receiver are not

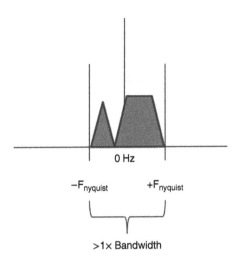

Figure 3.13 Complex-Valued ADC Bandwidth

necessary. This makes the DCR less expensive. There are multiple complications involved with direct conversion. Those shortcomings are discussed in [8]. Design considerations for direct-conversion transceivers are discussed in [9]. Because direct conversion is of particular interest to IoT, some of those shortcomings will be discussed here, including "IQ imbalance," "DC Offset," and "LO Leakage."

"IQ Imbalance" is caused by imperfections between the analog components in the I and Q arms, resulting in gain and phase imbalance. Consider that the I and Q arms independently carry a real-valued signal, meaning that the signal has both positive and negative (conjugate) frequency components. If the I and Q arms are 90-degrees out of phase and perfectly balanced in gain then the undesired frequency components will cancel out when summed as a complex value. However, if the phase and gain is not perfectly balanced, then there will be a residual undesired frequency component. The effects of IQ imbalance are graphed in Figure 3.14. An analytic equation for the relationship between the suppression of the undesired frequency components and the imbalances in gain and phase is provided in equation (3.4). This can be used to predict the Spurious-Free Dynamic Range (SFDR) afforded by the DCR. For example, in order to provide −50 dBc SFDR, gain imbalance must be less than 0.2 dB and phase imbalance must be less than 1 degree. G stands for gain imbalance, calculated in equation (3.3), and is applied to linear amplitude in equation (3.4).

$$G = 10^{\left(\frac{Imbalance\ (dB)}{20} \right)} \tag{3.3}$$

$$Suppression\ (dBc) = 10\log_{10}\left(\frac{G^2 - 2G\cos\varphi + 1}{G^2 + 2G\cos\varphi + 1} \right) \tag{3.4}$$

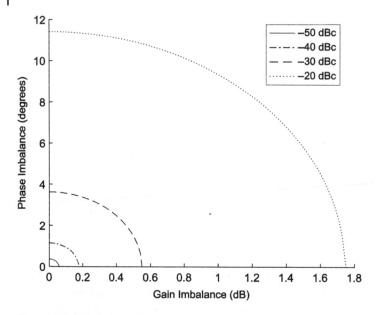

Figure 3.14 IQ Imbalance Curves

A difference in gain of 0.5 dB may seem very small to the casual reader. However, this seemingly small mismatch results in extremely degraded performance. An analytic expression for the relationship between IQ imbalance and an increase to the symbol error rate is detailed in [10].

"DC Offset" occurs during digitization. There are two ADCs, and the ADCs will have different offsets. The ideal ADC provides a linear relationship between input voltage and output "quanta" or "counts," those being the digital representation of the amplitude of the input signal. Unfortunately, there is no ideal ADC. The transfer characteristic of a real ADC, plotting input voltage to output counts, does not cross through the origin. That y-intercept point is the "DC offset." The two ADCs will have different DC offsets. A DC offset exists in all three of the architectures described in this section. The fact that the direct-conversion architecture operates at baseband with DC in the middle complicates the problem. Where the superheterodyne architecture operates at an IF bandwidth and can simply filter out any undesired residue at or near DC, the direct-conversion architecture must mitigate the imperfection. Furthermore, the direct-conversion architecture has two different DC-offsets, which can complicate algorithms used to correct the offsets. Digital-to-Analog converters (DACs) suffer the same issue. Again, where superheterodyne transmitters would simply filter out residue not centered at IF, the direct-conversion transmitter will forward the DC offset as a signal into the analog up-conversion

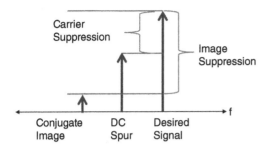

Figure 3.15 Simplified Direct-Conversion Spectrum

stages. This DC offset provides a 0 Hz component for the up-converting mixers to mix with.

"LO Leakage" occurs during complex-valued mixing. It must be remembered that analog mixers only emulate multiplication. There are undesirable products that result from analog mixers. Among these is the fact that some power from the LO will bleed through the analog mixer and evidence itself in the mixed output. For direct-conversion transmitters, this means some power from the Local Oscillator will be present in the middle of the transmitted bandwidth of interest.

There are other artifacts at baseband that pollute the bandwidth of interest. These include "flicker noise" from amplifiers and "dithering" at digitization stages.

As shown, there are a number of technological limitations that prevent the DCR from perfectly emulating Euler's formula. Ultimately, the user can expect the bandwidth of interest of direct-conversion transceivers to yield a smaller SFDR, therefore causing more distortion than that of a superhet. The effects of these imperfections are illustrated in Figure 3.15. Three signals are shown, the desired signal, the spur at DC, and the conjugate image. The conjugate image is caused by IQ imbalance. The spur at DC is caused by a number of factors, including LO leakage and DC Biases at the ADCs.

The undesirable effects due to the imperfections of analog components can be mitigated in software using algorithms such as those found in [11]. When paired with digital channelization, discussed in Section 3.2.2, these problems can be further mitigated by "off-tuning" the DCR by a frequency equal to at least one-half of the bandwidth of interest.

Since direct-conversion architecture pollutes the center of the bandwidth of interest, as it is at baseband, many wireless standards leave that region blank. For example, the OFDM symbol for 802.11 leaves the 0 Hz bin as "null," meaning there is no information there. This concept is shown in Figure 3.16. If the direct-conversion front-end can be designed to constrain some of the undesirable artifacts to within that bin, then the direct-conversion receiver can be

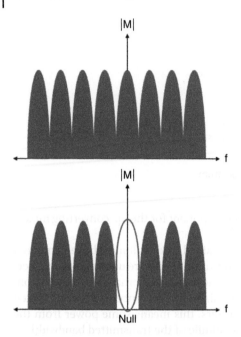

Figure 3.16 Null at Center of OFDM Symbol

free of those limitations. This leaves only IQ imbalance as the main detraction to using the direct-conversion architectures for commercial IoT applications. Applications like DVB-T, though not an IoT system, do populate the center bin and this implies that designers of the standard expect consumer electronics to not utilize a direct-conversion front-end.

3.2.1.3 RF-Sampling

The concept of RF-Sampling is simply to run the ADC at a clock rate high enough to sample the entire received spectrum. A pre-selection filter prevents out-of-band interference. The pre-selection filter may not need to be a bandpass filter, depending on the application. A Low Noise Amplifier (LNA) provides gain at the front of the receiver. There is no tuning stage in an RF-sampling architecture. The RF signal is directly sampled by the ADC. The same holds for RF-sampling transmitters. The output of the DAC, after anti-image filtering, is fed into an amplifier and transmitted.

As can be seen in Figure 3.17, the RF-sampling architecture has no tuning stage. This fact makes it very useful for operations in the high frequency (HF, 3 to 30 MHz) band. RF-sampling architectures are commonplace for applications in the HF band or lower. The fact that no tuning stage is used does not negate the need for frequency correction in the modem. Frequency correction, as part of synchronization, is discussed in Chapter 4.

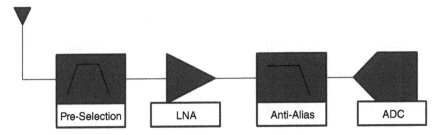

Figure 3.17 RF-Sampling Receiver

3.2.1.4 Summary

The pros and cons of the three RF front-end architectures discussed are summarized in Table 3.1. Superhet architecture can tune to a wide range of frequencies and provides a very strong SFDR. The superhet is also the most expensive. Direct-conversion architecture also tunes to a wide range of frequencies like the superhet and is inexpensive relative to the superhet. Direct-conversion architecture offers twice the bandwidth as the superhet. However, direct-conversion architecture suffers from issues as undesirable baseband artifacts of mixing, amplification, and digitization leak through into the bandwidth of interest. RF-sampling architecture is also inexpensive as compared to the superhet, but is obviously limited in tuning capability.

The shortcomings of direct-conversion architecture can be mitigated by way of algorithmic corrections and also by way of allowances in the wireless standard. As discussed in Section 3.2.1.2, the fact that the center bin in OFDM systems is left null allows inexpensive direct-conversion architectures to be used without the signal suffering from undesirable baseband artifacts of analog processes.

3.2.2 Digital Channelization

An optional component in the transceiver is digital channelization. Digital channelization is used if the analog front-end is sufficiently "wideband" to

Table 3.1 Analog Front-End Summary

Type	Pros	Cons
Superhet	Low spurs, wide frequency range	Expensive components
Direct conversion	Wide frequency range, double bandwidth	High spurs
RF sampling	Simple design, low spurs	Limited frequency range

require additional selectivity. The terms "wideband" and "narrowband" can mean many things in different contexts. For the purposes of this discussion, "wideband" means that more than one signal has been digitized, and "narrowband" means only one signal is digitized.

Digital channelization is common to software-defined radio (SDR) platforms. This is because most SDRs are wideband. The SDR being wideband allows the SDR to maximize re-configurability.

Excellent analysis of digital channelization is provided in [12]. The reader is encouraged to delve more deeply into the components and design of digital channelization. This section will provide an overview of digital channelization. This overview is intended to present the reader with an operational understanding of digital channelization as the second stage of channelization in the RX and TX chains of their transceiver.

Digital channelization is of interest to IoT applications for the simplifications it brings to transceiver design. Digital channelization can be used to implement a frequency-hopping solution without the need for fast retuning times in analog hardware. Frequency-hopping is discussed in greater detail as part of modems in Chapter 4. If the entire bandwidth of interest is digitized, digital channelization can be used to select a signal and quickly hop along with it. This works so long as one operates in a band where the front-end will not be overpowered and saturated by ambient signals. If the band of operation contains such risk, a narrowband solution would be preferable. Given that IoT applications tend to operate in bands with expressly limited transmit power, wideband front-ends are a good choice.

When the transceiver must perform additional channelization operations after digitization, the process is referred to as "digital channelization." The term can be applied to a variety of multiple access schemes, including: time-division multiple access (TDMA, discussed in Chapter 5), frequency-division multiple access (FDMA, discussed in Chapter 5), and code-division multiple access (CDMA, discussed in Chapter 5). Many wireless standards will employ two or more multiple access schemes concurrently. A common example is a system comprised of multiple TDMA signals separated into different frequency channels (FDMA). A transceiver for such a system may digitize the entire operational band and then select frequency channels of interest by way of digital channelization.

It is common to employ Digital Down-Converters (DDCs) and Digital Up-Converters (DUCs) for FDMA systems. DDCs and DUCs may be implemented in separate chips within RX and TX chains, respectively. The data rate employed at the digitization stage of a wideband receiver is generally too massive for baseband controllers to process. The channelization of this large data rate to a smaller, more selective data rate is pushed onto other devices. There are DDCs and DUCs available as separate Commercial-Off-The-Shelf (COTS) components, or the components may be instantiated on an FPGA. This is an example of the processing on transceiver hardware being heterogeneous.

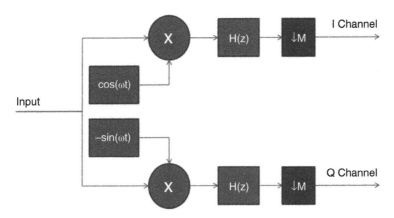

Figure 3.18 Digital Down-Converter Structure

A simplified block diagram of DDC is illustrated in Figure 3.18. The DDC utilizes two phase-orthogonal (90 degrees out of phase) sinusoids to emulate Euler's Formula, as in equation (3.2). In fact, DDC is a direct-conversion structure. The difference between DDC and direct conversion architecture, described in Section 3.2.1.2, is that DDC is digital and does not suffer from any of the impairments that analog components inflict on analog direct-conversion front-end architecture. There is no IQ imbalance, DC offsets, LO leakage, or other analog domain impairment. The operation of DDC is shown in Figure 3.19. The DDC uses locally generated sinusoids to down-convert a signal to baseband. The DDC then isolates that selected signal and reduces the sample rate. The fact that the sample rate has been reduced is demonstrated by the compression of the Nyquist bandwidth.

DDCs may be employed to select smaller bandwidths from wideband front-ends. This digital sage of channelization enables the off-tuning mitigation strategy. In this strategy, a DCR can tune the polluted center of its band away from the bandwidth of interest. The center frequency of the DCR is tuned away from the desired frequency by at least one-half of the bandwidth of interest. A downstream DDC then down-converts the bandwidth of interest to baseband, filtering out the pollution at DC and the undesirable conjugate image at the same time.

The DDC implementation in Figure 3.18 converts a real-valued bandpass signal into a complex-valued baseband signal. This accommodates RF-sampling or superheterodyne front-ends. In order to accommodate a direct-conversion front-end, the digital mixing stage shown in Figure 3.18 must change to that shown in Figure 3.20. The negative sign on the sine makes the process downconversion. If the sine were positive, the frequency conversion would be up-conversion.

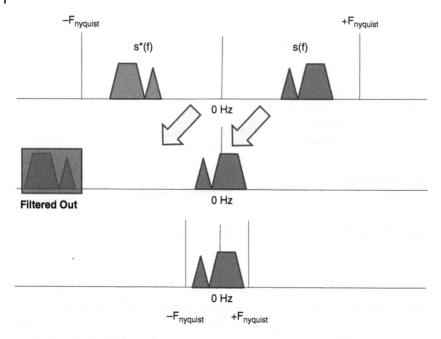

Figure 3.19 DDC Operation

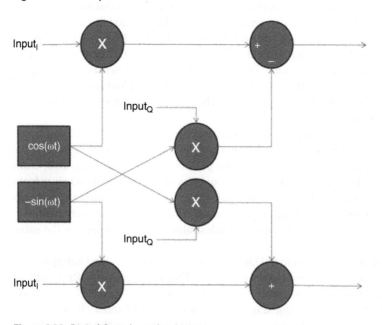

Figure 3.20 Digital Complex-Valued Mixing

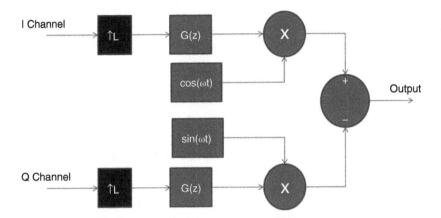

Figure 3.21 Digital Up-Converter Structure

Digital Up-Conversion is the process whereby, in the digital domain, a complex-valued signal at baseband is up-converted to a wider Nyquist bandwidth. A simplified block diagram of DUC is illustrated in Figure 3.21. A complex-valued baseband signal is brought into the DUC. The baseband signal is interpolated to a higher sample rate. The baseband signal is then projected to a new frequency using phase-orthogonal sinusoids and Euler's formula. The operation of the DUC is illustrated in Figure 3.22. In that example, a

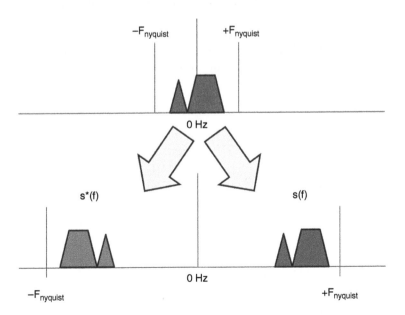

Figure 3.22 DUC Operation

complex-valued baseband signal is converted to a real-valued bandpass signal. DUC can be used with a direct-conversion transmitter. The mixing stage must be modified, as discussed for DDC and as shown in Figure 3.20, except that the sine must be positive.

3.2.3 Baseband Controller

The baseband controller controls all functions of RX and TX chains. The baseband controller sets the center frequency, gains (or set-points for automatic gain control), channelization, and other aspects of the processes required to receive and transmit signals. The baseband controller is analogous to the "baseband processor" common in cellphone applications. The term "baseband controller" is more common for IoT applications and the term "baseband processor" is more common for cellphone applications. The differences vary between manufacturers, applications, and other contexts. The baseband controller may be a Commercial-Off-The-Shelf (COTS) device or it may be a processor running user-defined software.

The baseband controller may process signals at baseband, thus implementing the modulation and demodulation (MODEM) of the desired signal. An example of such a system is given in [13]. Such a device is used in a heterogeneous system by sending demodulated data to another device dedicated to higher layers of the protocol stack and taking data from that device to be modulated and transmitted. Heterogeneous processing is common in modern communication system architectures. Heterogeneous processing may be contained within one System-on-a-Chip (SoC).

Depending on the design, the baseband processor may connect directly to the analog front-end, thus bypassing the need for digital channelization. Such a configuration is shown in Figure 3.23. In this example, the analog front-end

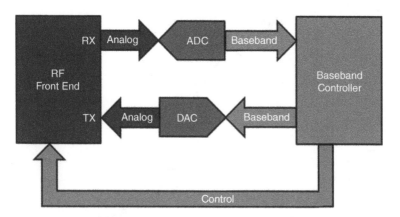

Figure 3.23 Narrowband System

is narrow-band enough to preclude the need for further channelization. The baseband controller simply absorbs the whole digital signal.

3.3 The Basics of Channels

The channel is the one portion of the wireless communication system that developers cannot design. It is the physical world and the environment in which the system is to be used that dictates the behavior of the channel. The developers of a wireless communication system can model a channel. The developers can take empirical data and compare that to their models. It is, therefore, important to build a foundation by which channels can be modeled such that the receivers can mitigate the negative effects imparted by the wireless channel. This section will provide definitions and models for the channel component of a wireless system.

3.3.1 What is a Channel?

When building a foundation for describing a wireless system, one of the first questions that should be asked is "what is a channel?" What separates one channel from another? If one tunes a broadcast radio receiver, has one actually changed the "channel" or has one simply "changed the frequency?" The answer may be "both." The term "channel" is defined almost entirely in context. That is important to remember. In order to disambiguate the term, a variety of qualifiers are appended to the word channel. Textbooks and wireless standards will employ these qualifiers to refer to different types of channels, such as "physical channels," "logical channels," "frequency channels," and many more.

A "physical channel" describes the medium through which a signal propagates. Physical channels will impart some effect on the data passing through it. Usually that effect is undesirable. There are several models for physical channels, which shall be discussed in a later section.

A "logical channel" is the selection of some subset of data from a composite signal. The concept of a logical channel is related to the concept of multiplexing. If a single signal contained data for ten different users, the logical channels for that signal would differentiate between the data intended for each individual user. Multiplexing will be discussed in detail in a later section.

A "frequency channel" is a predefined frequency, which, when selected, will provide a desired signal. This means that a "frequency channel," sometimes called a "broadcast channel," can be described as both a "physical" and "logical" channel. Examples of frequency channels are radio stations. One tunes their radio receiver to the radio station frequency and in doing so, selects their desired broadcasting and also recovers a desired signal from a physical channel. Another example of frequency channels is the Absolute Radio-Frequency

Channel Numbers (ARFCN) of GSM, now subsumed as part of the Universal Mobile Telecommunication System (UMTS). UMTS restricts the term "logical channel" to further channel selection after frequency tuning, but that fact demonstrates the contextual nature of the meaning of the word "channel."

3.3.2 Simple Physical Channel Models

There are several physical channel models. Those models are not necessarily exclusive. Different physical channel models can be combined to form a more robust model of the wireless link. This section shall introduce several models independently and then combine those models at the end.

3.3.2.1 The Large-Scale Fading Channel and Link Budget Analysis

This section will describe large-scale fading and link budgets. A link budget analysis is an attempt to determine the loss in power a signal will suffer when transmitted to a receiver some distance away.

The link budget determines the received power of a wireless link, taking into account the transmit power, antenna gains, propagation losses, and a margin for channel effects. It is important to guarantee a minimum signal power at the receiver. The link margin takes this into account and allows for various channel effects that will reduce the received signal power beyond propagation loss.

Fading is a physical phenomenon that impairs the power of a transmission. "Large-scale" fading represents attenuation that impairs the transmission consistently over long distances. Path loss is accounted for in large-scale fading. Equation (3.5) shows the relationship between path loss and distance. The path loss is proportional to relative distance raised to an exponent, n. The exponent n is specific to the operational environment. For a Line-Of-Sight (LOS) link, the exponent is typically 2. Standards and other resources will provide exponents for Non-LOS (NLOS) links based upon field measurements taken in different types of environments. A common exponent to use for NLOS is 4.

$$L_p \propto d^n \tag{3.5}$$

A simplified link budget follows the Friis Transmission equation:

$$P_r = P_t + G_t + G_r - L_p \tag{3.6}$$

Where   Pr is the received power in decibels relative to a reference (dBm, dBW, etc.),

          Pt is the received power in decibels relative to a reference (dBm, dBW, etc.),

          Gt is the transmit antenna gain in dBi,

          Gr is the receive antenna gain in dBi,

          and Lp is the path loss in dB.

Antenna gain is directly proportional to antenna directivity. The directivity of an antenna is related to the beamwidth of an antenna. Gain is increased by concentrating the radiated power in one preferred direction. Therefore, an antenna with high gain must be pointed at its intended target. The relationship between antenna gain and antenna directivity can be expressed as:

$$G = \varepsilon D \qquad (3.7)$$

Where   D is the directivity of the antenna
and ε is the efficiency of the antenna.

The gain is measured in decibels relative to an isotropic radiator. An isotropic radiator is an ideal antenna that exists only in theory. The isotropic antenna radiates power uniformly in all directions. An antenna with gain does not.

The use of directional antennas in the design of a wireless system is called sectorization. For example, a transmitting basestation may employ three 120-degree directional antennas. This gives the basestation the ability to cover 360 degrees in azimuth and provide a gain for the wireless link. These antennas only cover 360 degrees in azimuth; the elevation angle is a separate concern.

Omnidirectional antennas are antennas that transmit uniformly in azimuth, but not in elevation. The antenna pattern forms a donut-like shape. This may seem a misnomer because the "omnidirectional" antenna is limited in elevation angle beamwidth. The elevation angle covered by any one omnidirectional antenna varies. As described above, the smaller the elevation angle covered, the larger the gain of the omnidirectional antennas. A low gain omnidirectional antenna may have a gain of 2 dBi. This allows for a very large elevation angle to be covered.

Antenna height also matters in the ability of one antenna to illuminate another antenna. Obstacles can cause additional attenuation of the signal by way of antenna geometry. If the signal needs to pass through an object to reach the intended receive antenna, that signal will suffer some non-negligible attenuation. To keep this example simple, such specifics in the physical layout of the system will be bypassed.

For link budget analysis, the path loss can be computed as follows:

$$L_p = n * 10\log_{10}(d) + 20\log_{10}\left(\frac{4\pi}{\lambda}\right) \qquad (3.8)$$

Where λ is the wavelength of the wireless signal,
d is the distance between the transmitters,
and n is the large scale fading exponent.

Wavelength is a factor in path loss because the wavelength determines the size of the antenna. Larger antennas provide more area to be illuminated by the power of the transmitter. Smaller antennas for smaller wavelengths (higher frequencies) have a smaller footprint.

Note that the loss due to antenna size as dictated by the wavelength is not affected by the large-scale fading exponent. The loss due to antenna size is also not affected by distance.

After taking into account the large-scale fading, the link budget needs to provide margins for other effects. These margins are margins of error that we will reserve in our link budget for other channel impairments. These effects include physical phenomena, such as log-normal shadowing and multipath. These other impairments will be discussed later. The transmission equation now becomes:

$$P_r = P_t + G_t + G_r - L_p - L_m \qquad (3.9)$$

Where L_m is the expected loss due to a stochastic channel impairment.

Let's try an example calculation. The Bluetooth standard specifies a reference receiver sensitivity of −70 dBm. Receiver sensitivity is defined by the standards as the signal strength (in power) at the receiver. The standards require that at this minimum, the receiver be able to achieve a Bit Error Rate (BER) of one error in 1000 bits. A class 3 transmitter transmits at 0 dBm. As we will see, this is a very gracious requirement. It is clear that the standard was created with the intention of employing very inexpensive components and subsequently low performance in terms of radio hardware.

Both transmit and receive antennas are omnidirectional with a gain of 2 dBi. The link margin is chosen as 10 dB. The band of operation is the 2.4 GHz ISM band. The link is simple Line-Of-Sight (LOS). What is the maximum distance of this link given that the minimum receive power is −70 dBm?

First, solve for the maximum allowable path loss.

$$L_p = P_t - P_r - L_m + G_t + G_r \qquad (3.10)$$

Transmit Power	0 dBm
Minimum Receive Power	−70 dBm
Link Margin	10 dB
Tx Antenna Gain	2 dBi
Rx Antenna Gain	2 dBi
Maximum Allowable Path Loss	64 dB

The minimum receive power is −70 dBm; however, the link margin of 10 dB increases this to −60 dBm. The gains of the receive and transmit antenna provide a boost, allowing the receive power to be −64 dBm. This means that path loss must be limited to 64 dB.

Now that the maximum allowable path loss is known to be 64 dB, we can relate that to distance. To do so, use equation (3.8). There are two summed components to path loss by way of the definition in equation (3.8). One component is a function of distance; the other is a function of wavelength. Remove the component of path loss determined by wavelength.

$$20\log_{10}\left(\frac{4\pi}{\lambda}\right)$$

Frequency	2.40E + 09 Hz
c	3.00E + 08 m/s
Wavelength	0.125 meters
Loss	40.05 dB

As can be seen, the band of operation (2.4 GHz) causes a loss of 40 dB due to the small size of the antenna. This means path loss due to distance must be limited to 24 dB.

$$n * 10\log_{10}(d)$$

The link is LOS and thus the large-scale fading exponent is 2. Solving for distance yields a maximum distance of 15.84 meters. If the link was NLOS, using large-scale fading exponent of 4, then the maximum distance would be 3.981 meters.

Both of the maximum ranges calculated above extend well beyond the expected operational range. Expected operational ranges for Bluetooth device power classes are listed in Table 3.2. Class 3 devices are only expected to operate within 1 meter. The class 3 operational range falls well short of 15.84 meters. So what are the losses for which the calculation above has not accounted? It is difficult to impose an obstruction, like a brick wall, between the receiver and transmitter in this short operational range of 1 meter. What's more, Table 3.2 demonstrates an expected linear relationship between distance and power.

Table 3.2 Bluetooth Power Classes and Operational Ranges

Class	Power	Operational Range
Class 1	100 mW	100 meters
Class 2	2.5 mW	10 meters
Class 3	1 mW	1 meter

There is no case where the exponent in equation (3.5) is 1. Yet, there is a clear expectation that when the power increases to 100 mW, the range will increase 100-fold.

What this represents is the expectation that class 3 devices and systems will be very inexpensive but suffer impairments due to using very inexpensive components. For example, the antennas used may be highly inefficient. The demodulators may require a very high SNR in order to achieve the expected bit error rate. By contrast, we can see that class 1 devices are expected to use far less lossy components.

3.3.2.2 The Additive White Gaussian Noise Channel

The large-scale fading model provides a model for the loss of power in the transmitted signal as that transmitted signal propagates in distance. But why should that matter? What prevents a receiver from recovering a very weak signal? The answer is, primarily, noise. It is noise in the receiver that prevents the receiver from recovering weak signals.

The Additive White Gaussian Noise (AWGN) channel models noise inherent to communication systems due to thermal energy. Each term in the name has a specific meaning.

Additive: Because the noise signal is specifically added to the received signal, as opposed to being convolved or multiplied with the received signal, the noise signal is described by the term "additive." This noise is modeled as a random signal added at the beginning of the receiver. The AWGN channel may be used independently to determine theoretic bit error rate performance. When used in conjunction with other channel models, the AWGN channel model must be placed directly before the receiver after the other models have modified the signal. Propagation loss does not diminish the noise signal created by the AWGN channel. AWGN does not model any effects at the transmitter. The AWGN channel models the noise of the receiver.

White: Each sample of the noise signal is independent of any other sample. Because each sample is independent, the autocorrelation of the noise signal results in an impulse function. The power spectral density of a random signal is the Fourier transform of the autocorrelation. Therefore, the Fourier transform of the autocorrelation of the noise signal is constant across all frequencies. The frequency domain representation of a random process is a power spectral density measured in units of power over hertz. Because the spectrum of noise will be flat, one can describe a level of the power spectral density, N_0. Noise power, as measured in units of power, can only be determined after a bandwidth has been defined. This is shown in equation (3.11). N_p is the noise power, B is the bandwidth of the receiver, and N_0 is the noise power spectral density. Without a defined bandwidth, noise power at the receiver is infinite [2].

$$N_p = BN_0 \qquad (3.11)$$

Gaussian: The noise signal is a result of a composite of independent random noise sources. This random composite signal of multiple small noise sources has a Gaussian distribution function by way of the "central limit theorem." A discussion on the physical causes exceeds the scope of this chapter. More information on the central limit theorem and physical noise sources can be found in [2].

This definition provides a description of a random signal that one can use to model the noise inherent in the receiver. The level of the power spectral density of that noise depends upon the source temperature and effective noise figure of the receiver. Noise figures and noise temperatures are discussed in great detail in [4] and [2].

3.3.2.3   Conclusion: The Simple Propagation Channel

Large-scale fading provides a model by which one can predict the loss of power in a transmitted signal over distance. AWGN in the receiver provides a model by which one can predict the amount of disruptive noise that the transmitted signal must overcome. The ratio of the power of the received signal to the power of the noise within the receiver bandwidth is SNR. The SNR directly relates to energy per bit (E_b) and noise power directly relates to the level of noise power spectral density (N_0). The ratio E_b/N_0 is used as a figure of merit for demodulator designs. Tests can be built around supplying a demodulator under test with a signal with a specific E_b/N_0 and then measuring the bit error rate. The bit error rate depends upon the specifics of a demodulator implementation.

There are two potential faults in the receiver design. The receiver hardware may be of too low quality to provide a sufficient E_b/N_0 to the demodulator, and the demodulator may be of too low quality to utilize the E_b/N_0 that the receiver hardware can provide. The link margin allows the designer to avoid costly components on each so long as the minimum bit error rate set by the standard is met.

Example: The Bluetooth standard specifies a minimum receiver sensitivity of −70 dBm. At this minimum, the standard states that the raw bit error rate shall be 0.1% or lower. Therefore, the receiver hardware must be able to receive a signal at −70 dBm and supply a digital signal with a sufficient E_b/N_0 to the demodulator such that the raw bit error rate is at or below 0.1%. The 0.1% raw BER indicates the minimum level of performance required to maintain a closed Bluetooth link. The term "raw bit error rate" means that error correction has not been employed.

The noise power can be calculated by knowing the receiver bandwidth and a value called "equivalent noise temperature" [2,4]. The details of this calculation are beyond the scope of this book. It is sufficient for this example to say that at normal ambient temperature (290 degrees Kelvin), the noise power spectral density will be −174 dBm/Hz. For this example, the hypothetical receiver with inexpensive components will add 26 dB to this spectral density, resulting in a noise floor of −148 dBm/Hz. For this hypothetical receiver, for the Bluetooth example we will set the receiver bandwidth suboptimally wide at 2 MHz.

N_0	−148 dBm/Hz
B	2 MHz
N_p	−85 dBm

If the signal is received at −70 dBm, as in the minimum receiver sensitivity example, then this provides a Signal-to-Noise Ratio (SNR) of 15 dB. The demodulator must be capable of demodulating the signal with one error in 1000 bits at this SNR. Is that reasonable?

To answer that question, we must look to the BER curve. BER curves show the probability of bit error given a ratio of energy per bit over noise power density (E_b/N_0). Note that the BER curve is not plotted against SNR. The received SNR must be converted to E_b/N_0.

3.4   Bit and Symbol Error Rate

There are numerous error rates discussed in the design of a wireless system. These include frame error rates, packet error rates, bit error rates, symbol error rates, and others. The bit error rate and the symbol error rate are the two error rates most commonly discussed. The bit error rate is the expected rate of errors in individual bits. The symbol error rate is the expected rate of errors in symbols parsed from the over-the-air waveform.

Symbol error rates are mathematically defined and experimentally measured as functions of the ratio E_s/N_0, which is the received energy per symbol (E_s) measured in Joules to the spectral noise density (N_0) measured in Watts/Hz. The unit of Watts/Hz is called "power spectral density" and can be thought of as another representation of a unit of energy.

Signal energy is defined as shown in equation (3.12). $E_x(t)$ represents the energy of some signal x at some time t. This is the integral of the instantaneous power of x over time. In this context, power is measured in standard units such as Watts. Energy would be measured in corresponding units such as Joules.

$$E_x(t) = \int_{-\infty}^{t} P_x(\tau)d\tau \qquad (3.12)$$

The energy in a given symbol is defined as shown in equation (3.13). $E_s[n]$ represents the energy of the nth symbol. The energy of the nth symbol is the integral of the power delivered during that interval of time. This time interval is based on the symbol period, T_s.

$$E_s[n] = \int_{(n-1)T_s}^{nT_s} P_x(\tau)\,d\tau \tag{3.13}$$

By way of equations (3.12) and (3.13), one would need to know the exact shape of the symbol and the instantaneous power delivered as a function of time. A quicker rule of thumb to determine the energy delivered in each symbol interval is to multiply the average power of the signal with the symbol period. This provides an approximation of the average energy delivered for each symbol. This product is shown in equation (3.14).

$$\overline{E_s} = \overline{P_s}T_s \tag{3.14}$$

By way of equation (3.14), we can quickly derive a value for energy per symbol based upon the symbol period and the received power. Given that the symbol period is the inverse of the symbol rate, R, then combining equations (3.14) and (3.11) yields equation (3.15), which shows a simplified rule-of-thumb relationship between SNR and E_s/N_0. The ratio of the receiver bandwidth divided by the symbol rate provides a scalar to be applied to SNR to convert that received value to E_s/N_0. E_s/N_0 can then be used to predict the symbol error rate.

$$\frac{\overline{E_s}}{N_0} = \frac{\overline{P_s}B}{N_pR} = \frac{B}{R}SNR \tag{3.15}$$

If the modulation order of the waveform is 2, meaning that the scheme can only transmit "1s and 0s," then the terms "bits" and "symbols" are synonymous. If, however, a higher-order modulation scheme is used, then the modulation scheme is transmitting multiple bits in each symbol. In the case of higher-order modulation schemes, symbol and bit error rates are different.

There is additional context and nuance to the discussion. If forward error correction is utilized, then a distinction may be drawn between "data bits" and "code bits." This is to say that a certain number of errors in "code bits" can be tolerated and result in no errors in "data bits." Many wireless standards, such as Bluetooth, specify a maximum bit error rate, and this criterion is meant to be interpreted as bits before Forward Error Correction (FEC) is applied. By specifying that we are discussing symbol error rate, such nuance is side-stepped and we explicitly describe errors resulting from the reception and demodulation of the wireless waveform.

The symbol error rate for a particular system wholly depends on the receiver implemented for that system. This is because there are multiple ways to demodulate a given signal. The symbol error rate is a function of the method chosen.

For example: A Binary Frequency Shift Keyed (BFSK) signal can be demodulated coherently or non-coherently. The non-coherent demodulation of BFSK can be based upon correlation or a frequency discriminator. This implementation choice will drive the error rate curve. This choice also drives cost. One need not choose the best error rate curve; one only needs to satisfy the requirements of the wireless link.

Margins were established in Section 3.2.2. These Margins account for lognormal shadowing and multipath and other impairments.

Suboptimal receiver design is not usually considered "an impairment" in the context of a link budget; however, poor performance can certainly be categorized as an impairment to the system as a whole. There is an important distinction here. The link budget allows for a guarantee that a minimum signal power will be at the receiver. This assumes that the receiver is capable of demodulating the signal at that received signal power and not suffer so high a bit error rate as to make the wireless link useless. However, if one intended to make a cost-effective but suboptimal receiver, one would raise the minimum receiver sensitivity to a level at which the SNR of the received signal would result in an E_b/N_0 sufficient for the suboptimal receiver to demodulate the received signal with a number of bit errors low enough to allow the link to be useful.

Picking up the example from the previous section, the signal is received at an SNR of 15 dB. For this example, the receiver bandwidth will be larger than necessary at twice the data rate. Bluetooth basic rate operates as two-level CPFSK with a modulation index of 0.315. These terms will be explained in Chapter 4.

Because the modulation scheme is two-level, the symbol energy is the bit energy. The E_b/N_0 of this received signal 12 dB as per equation (3.15) because the noise bandwidth of the receiver is twice as large as the data rate.

Figure 3.24 plots the BER curve for this modulation scheme in the presence of AWGN. This plot does not account for Gaussian pulse shaping, also discussed in Chapter 4. An E_b/N_0 of 12 dB gives us a bit error rate less than one error per 10,000 bits.

3.5 Complex Channels

Building upon the foundation laid in Section 3.3, this section will add stochastic elements to channel response. Section 3.3 describes large-scale fading due to propagation. In the succeeding sections, this propagation loss and the resultant received signal power is related to bit or symbol error rates with the addition of AWGN. In this section, we introduce more difficult concepts that occur in terrestrial channels. It is beyond the scope of this work to delve deeply into the propagation of electromagnetic waves across various terrain features and obstacles. References for the interested reader include [14] and [3].

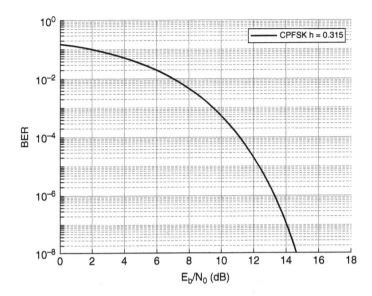

Figure 3.24 Bit Error Rate Curve for CPFSK h = 0.315

3.5.1 Shadowing and Large-Scale Fading

There are random perturbations in the large-scale fading of a signal over large distances. In order to account for these random perturbations, a stochastic term is added to equation (3.6) from Section 3.3.2.1. The random contributions to path loss in decibels by way of these perturbations follow a Gaussian, or "Normal," distribution. Because the random additional loss follows a normal distribution when that contribution is on a log-scale, this type of distribution is called a "Log-Normal" distribution. The extra loss is only a Gaussian random variable when added to path loss as decibels.

3.5.2 Small-Scale Fading and the Multipath Channel

An excellent description of small-scale fading can be found in [14]. From [14] comes Figure 3.25, which illustrates link budget planning with a stochastic model and details each contributing factor. This illustration shows how to plan out a link budget that plans for non-deterministic channel effects. Distance is represented on the x-axis and power is represented on the y-axis. Contributions to the loss of signal power are large-scale fading, near-worst case variations about the value of path loss, and near-worst-case Rayleigh small-scale fading.

The first effect is power loss due to propagation. As distance increases, power decreases. This path loss is deterministic.

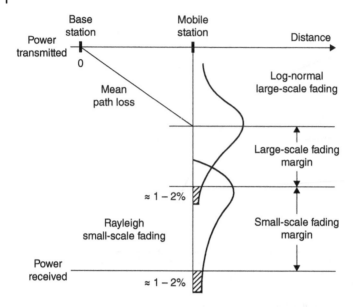

Figure 3.25 Link Budget Considerations for a Fading Channel [14]

The expected loss in power due to path loss provides a mean for non-deterministic large-scale loss due to shadowing. Shadowing is a log-normal random loss centered at the expected power level left after path loss.

Loss due to small-scale fading is added to the large-scale fading. The amount of margin indicated is intended to provide adequate received signal power for approximately 98–99% of each type of fading variation (large- and small-scale). To that end, the mean of the probability distribution of small-scale loss is placed at the lower tail-end (98% mark) of shadowing loss. The expected received power is then taken from the lower tail-end of small-scale loss. This worst case analysis allows the system designer to plan an adequate link budget.

Small-scale fading represents the changes in amplitude and phase of the received signal that are observed as a result of small changes in the positioning between a receiver and a transmitter. That is in contrast to large-scale fading that covers the effect over large distances. This section will discuss the small-scale fading phenomenon first with stationary nodes and then with moving nodes.

Large-scale fading and AWGN can be combined together to create a model of a channel in which transmission distance is limited to the SNR at the receiver necessary to retrieve bits from the transmitted signal. As discussed in the large-scale fading section, IoT link budgets usually establish a wide margin between the minimum needed SNR and the SNR received at an operational distance. This is done in part due to the environments where IoT will be used. These environments are often indoors. Indoor environments are "multipath" environments and will inflict a convolutional impairment on the received signal. Where AWGN is an additive impairment and large-scale fading represents a multiplicative (scalar) impairment, multipath channels introduce a convolutional impairment meaning that the effect of the channel is to add a filter with some memory in between the transmitter and the receiver.

A multipath environment is one where the transmitted signal finds multiple pathways to the receiver. It sometimes helps to think of RF transmissions as a form of light; hence, the transmitter seeks to "illuminate" the receive antenna. With that in mind, consider a house of mirrors at a carnival. The optical illusions caused by the reflections are disorienting and make one object appear several times in several places. This is a multipath environment, and your eyes are the receiver. The light from a single object has multiple paths to your eyes. Each path has a different phase and angle.

There are three basic mechanisms behind a multipath environment. Those are reflection, diffraction, and scattering.

- Reflection: A transmitted signal can reflect off of an object and be re-directed towards the receiver. This effect is like that of a mirror.
- Diffraction: Diffraction is where a large object impedes the propagation of the transmission and secondary waves form behind the obstruction.
- Scattering: Scattering is like reflection off of a rough mirror. When the transmitted signal bounces off a rough surface, the energy is spread out in all directions.

These three physical phenomena provide means for a single transmission to reach a receiver multiple times. Consider a link with one direct "line-of-sight" path and two reflective surfaces nearby. This example is shown in Figure 3.26.

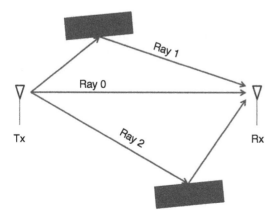

Figure 3.26 Three-Ray Multipath Example

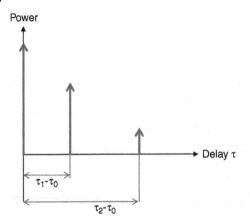

Figure 3.27 Three-Ray Multipath Example: Power Delay Profile

The paths emanating from the reflective surfaces travel a larger total distance than the direct path. These longer distances result in delays and loss in these additional paths. This is shown in Figure 3.27. Received power is plotted against time-delay τ. Time-delay τ represents the delay between the time the transmitter sent a signal and when that signal arrived at the receiver. For any transmission, there will be some delay in time as the signal traverses the distance between the transmitter and receiver at the speed of light. The first and largest power measurement in Figure 3.27 represents this familiar phenomenon of propagation delay and propagation loss. Sometime later, that same signal reappears at the receiver. These new appearances of the same signal are due to the multipath channel. Each of the additional paths delays and attenuates the signal separately from the main line-of-sight path. The delays and power of the additional paths are normalized to the main path.

A multipath environment can be modeled as a Finite Impulse Response (FIR) filter. The power delay profile can be considered coefficients in a "channel filter" modeled as an FIR filter. Thus, the channel has a channel impulse response. This is illustrated as a combined channel model in Figure 3.28. The channel has an impulse response h(t) and noise is added just before the receiver.

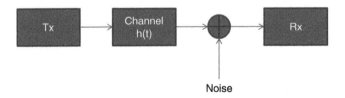

Figure 3.28 Combined Channel Model

The channel impulse response likely will "smear" the symbols, causing the symbol energy to exceed the symbol period and bleed into other symbols. This interference is called "Intersymbol Interference" (ISI). ISI is a serious concern for the receiver, and techniques are employed to resolve the issue. Some of this can be mitigated at the transmitter. If the transmitted waveform is not appropriately bandlimited, the channel impulse response may impose more undesirable effects. This will be discussed further in the next chapter.

Real-world environments for the wireless IoT include office buildings, apartment buildings, and factories. The wireless IoT operates in environments full of obstructions, reflections, diffraction, and scattering. A full analysis of the physics of any one environment is non-trivial. Furthermore, each environment is unique. A model for the channels of interest to the IoT is provided in [15]. Reference [16] goes into more details of channel modeling than can be covered in this chapter, and can be helpful when reading through the model outlined in [15].

References

1 R. G. Lyons, *Understanding Digital Signal Processing*. Upper Saddle River, NJ: Prentice Hall, 2010.
2 B. Sklar, *Digital Communications: Fundamentals and Applications*. Upper Saddle River, NJ: Prentice Hall, 2001.
3 T. S. Rappaport, *Wireless Communications: Principles and Practice*. Upper Saddle River, NJ: Prentice Hall, 2002.
4 D. Pozar, *Microwave Engineering*. Hoboken, NJ: John Wiley & Sons, 2005.
5 A. Parssinen, *Direct Conversion Receivers in Wide-Band Systems*. Boston: Kluwer Academic Publishers, 2001.
6 J. Tsui, *Digital Techniques for Wideband Receivers*. Raleigh, NC: SciTech Publishing, 2001.
7 E. H. Armstrong, "A new system of short wave amplification," *Proc. IRE*, vol. 9, no. 1, pp. 3–11, 1921.
8 A. Abidi, "Direct-conversion radio transceivers for digital communications," *IEEE J. Solid-State Circ.*, vol. 30, no. 12, pp. 1399–1410, 1995.
9 B. Razavi, "Design considerations for direct-conversion receivers," *IEEE Trans. Circuits Syst. II, Analog Digit. Signal Process.*, vol. 44, no. 6, pp. 428–435, 1997.
10 M. Windisch and G. Fettweis, "On the impact of I/Q imbalance in multi-carrier systems," in *IEEE Int. Symp. Circuits Syst.*, New Orleans, LA, May 2007, pp. 33–36.
11 S. Ellingson, "Correcting I-Q Imbalance in Direct Conversion," The Ohio State University, ElectroScience Laboratory, 2003.
12 J. H. Reed, *Software Radio: A Modern Approach to Radio Engineering*. Upper Saddle River, NJ: Prentice Hall, 2002.

13 K. Chen and H. Ma, "A low power ZigBee baseband processor," in *Int. SoC Des. Conf.*, Busan, South Korea, Nov. 2008, pp. I-40–I-43.

14 B. Sklar, "Rayleigh fading channels in mobile digital communications systems. Part I: Characterisation," *IEEE Commun. Mag.*, vol. 35, no. 9, pp. 136–146, 1997.

15 A. Molisch, K. Balakrishnan, C. Chong, S. Emami, A. Fort, J. Karedal, J. Kunisch, H. Schantz, U. Schuster, and K. Siwiak, IEEE 802.15.4a Channel Model - Final Report, 2004.

16 F. P. Fontan and P. M. Espineira, *Modeling the Wireless Propagation Channel: A Simulation Approach with MATLAB*, John Wiley & Sons Ltd., 2008.

4

Modem Layer

When examining wireless standards, it is important to keep in mind that the authors of the standard actually intended the standard to be used. It is intended that the developer succeeds in developing devices that can transmit and receive data according to said standard. The standard may appear to be obtuse, but that is only because the standard is not written as a set of instructions for the developer. The standard is written in such a way as to derive necessary tests and impose minimum limits to the developer's freedom. It is not the role of the standard to specify a singular implementation. The wireless standard will leave room for competitors to innovate.

Wireless standards are developed with economics in mind. Wireless standards often allow leeway for quality such that the developer can meet their market's needs. For example, in a cellphone network, it is often beneficial to make the user equipment (handset) inexpensive by defraying the complexity of the wireless system onto centralized basestation equipment.

Because the wireless Internet of Things (IoT) is based on digital data, the wireless IoT relies on digital modulation schemes. These digital modulation schemes require the physical layer (PHY) of a wireless IoT standard to utilize digital signal processing (DSP) in order to implement Modulation and Demodulation (MODEM) of the waveform.

DSP implementations are all too often esoteric and unapproachable. The esoteric nature of DSP implementations can prevent a thorough understanding of the reasons behind the requirements imposed by wireless standards. By examining the concepts in the standards from first principle, one can gain a better understanding of the concepts behind those standards and find new opportunities for innovation.

This book traverses the unified protocol stack model presented in Chapter 1. That stack is shown again in Figure 4.1. The placement of this chapter in that stack is illustrated by a lack of shading and an arrow.

In this book, the PHY is broken into two chapters, Radio and MODEM. The Radio chapter focused on physical concepts in communication theory as those

The Wireless Internet of Things: A Guide to the Lower Layers, First Edition. Daniel Chew.
© 2019 by The Institute of Electrical and Electronic Engineers, Inc. Published 2019 by John Wiley & Sons, Inc.

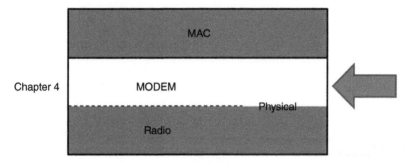

Figure 4.1 Traversing the Stack: The MODEM Layer

concepts applied to wireless IoT standards. This chapter will endeavor to shed light on modulation and demodulation. It is beyond the scope of one chapter, or even one book, to cover the entire breadth and depth of physical layer considerations for wireless communications. This chapter will discuss the various waveforms employed by wireless IoT standards. This chapter will detail the advantages and disadvantages of different waveforms, and how the selection of those waveforms is often driven by the desired application and transceiver hardware. Because no one chapter, or one book, can cover all concepts relevant to modem design, this chapter will provide a list of useful references for further study.

4.1 The Signal Model

In this section, the model of a complex-valued baseband signal including noise will be built. That model will start with the representation of the signal, continue to the establishment of the characteristics of noise in the complex plane and conclude with a unified expression of both.

4.1.1 Complex-Valued Signals

Due to the benefits of complex-valued representation, as described in Chapter 3, the reader is strongly encouraged to embrace the complex-valued signal model. An analytic expression for that model is given in equation (4.1).

$$s(t) = M(t)e^{j\theta(t)} \tag{4.1}$$

The magnitude and phase of the complex-valued signal are time-varying and possibly independent. The magnitude, $M(t)$, is always a positive value. Amplitude is the signed version of magnitude. When amplitude changes sign, magnitude does not; however, phase is inverted by 180 degrees. This fact will become useful when linear modulation is discussed in Section 4.3.1. For now, the information in the time-varying magnitude and time-varying phase need not be addressed.

The model shown in equation (4.1) can also be expressed as separable real and imaginary components. This is shown in equation (4.2). The real component is the "in-phase" component subscripted as "I." The imaginary component is the "quadrature" component subscripted as "Q." The real and imaginary components of equation (4.2) relate back to the phase and magnitude shown in equation (4.1). These relationships are shown in equations (4.3) and (4.4). "Arctan" is arcus tangent function and is used to generate the angle of the complex value. This instantaneous frequency of the complex-valued signal is the time derivative of the instantaneous phase, as shown in equation (4.5).

$$s(t) = s_I(t) + js_Q(t) \tag{4.2}$$

$$M(t) = \sqrt{s_I^2(t) + s_Q^2(t)} \tag{4.3}$$

$$\theta(t) = \arctan\left(\frac{s_Q(t)}{s_I(t)}\right) \tag{4.4}$$

$$\omega(t) = \frac{d\theta(t)}{dt} \tag{4.5}$$

As discussed in Chapter 3, the spectrum of a real-valued signal exhibits "conjugate symmetry," meaning that for each frequency component in the positive frequency range, there is a complex conjugate component in the negative frequency range. This concept is illustrated in Figure 4.2. Because the information in the negative frequency range is not unique and is, in fact, redundant, only the positive frequency range of the real-valued signal spectrum needs to be examined. A spectrum showing only the positive frequency range is called a one-sided spectrum. The spectrum of a complex-valued signal is different from that of a real-valued signal. The complex-valued signal has no conjugate images meaning the positive and negative frequency ranges hold unique information. Therefore, both the positive and negative frequency ranges must be examined for a complex-valued signal. A spectrum showing both positive and negative frequency ranges is called a two-sided spectrum.

The Nyquist Sampling theorem, briefly discussed in Chapter 3, imposes an upper limit on the signal bandwidth to be sampled without aliasing at a given sampling rate. By way of this theorem, no frequency component of the signal may exceed one-half the sampling rate in either the positive or the negative direction if aliasing is to be avoided. This limit of one-half the sampling rate is marked as the "Nyquist frequency" and the bandwidth defined between the

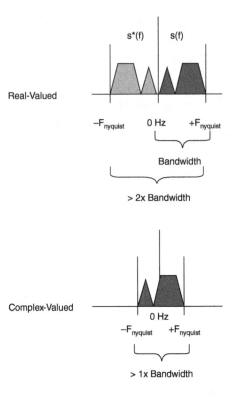

Figure 4.2 Spectrums of Real-Valued and Complex-Valued Signals

positive and negative Nyquist frequencies is equal to the sampling rate. Both the positive and negative frequency components of a signal must fit within this limit if aliasing is to be avoided. Therefore, a real-valued signal is limited to a bandwidth less than the Nyquist frequency, because a real-valued signal wastes half of its two-sided spectrum on conjugate images. By contrast, a complex-valued signal makes unique use of the two frequency ranges and is therefore limited to a bandwidth that is less than the sampling rate.

The complex-valued signal at baseband (around 0 Hz) can show the "constellation" of a signal. A constellation is a finite set of points on the complex plane at which the signal will be set in order to convey digital data mapped to that point. This is discussed in more detail in Section 4.3. The signal will be at baseband when addressing the constellation.

Complex-valued signals may exist either at baseband or at some intermediate frequency. Intermediate frequencies were discussed in the Radio chapter. A complex-valued signal may be deliberately placed at an intermediate frequency for a variety of reasons, including "off-tuning," as discussed in the Radio chapter. In addition to being deliberately offset from 0 Hz, the signal may have an inadvertent offset due to a frequency mismatch between the transmitter and the receiver. Such a frequency offset or mismatch can be corrected by carrier synchronization, which is discussed in Section 4.4.3.

4.1.2 Complex-Valued Noise

For complex-valued noise, there is an independent Gaussian noise variable on the I-arm and an independent Gaussian noise variable on the Q-arm. And expression for this is given in equation (4.6).

$$n(t) = n_I(t) + jn_Q(t) \tag{4.6}$$

As discussed in Chapter 3, by way of the Central Limit Theorem, the I and Q components have Gaussian distributions. To find the composite magnitude, the Pythagorean Theorem is employed. The I and Q components are squared then summed and then the square root of that sum is taken. This composite magnitude follows what is called a Rayleigh Distribution. A Rayleigh distribution is the square root of the sum of two Gaussian variables squared, as shown in equation (4.7).

$$|n(t)| = \sqrt{n_I^2(t) + n_Q^2(t)} \tag{4.7}$$

The noise signal now has phase, and that phase is important to the effects the noise imparts on signals. The composite phase is found by dividing the Q component by the I component and then taking the arctan of that ratio. The composite phase follows a uniform distribution. A collection of complex-valued noise samples are plotted in Figure 4.3. It can be seen that the phase for Rayleigh distributed noise is distributed uniformly.

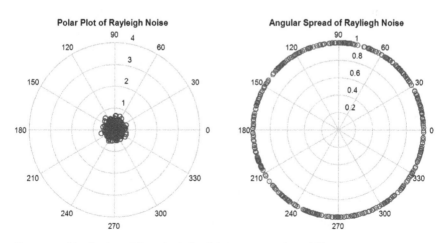

Figure 4.3 Distribution of Complex-Valued Noise Magnitude and Phase

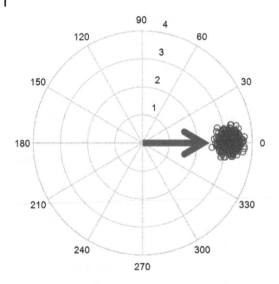

Figure 4.4 Complex-Valued Noise Moving with Increasing CNR

4.1.3   The Combined Signal Model

When a signal is added to complex-valued noise, as described in Section 4.1.2, the distribution of the magnitude of that sum changes from a Rayleigh distribution to a Rician distribution [1]. The total received signal, r, is composed of the desired signal, s, and the noise, n. As the signal becomes more powerful, the noise is pushed further away from the center of the complex plane, as shown in Figure 4.4.

The result is that the summed received signal now follows the Rician distribution. A collection of complex-valued samples of a signal in noise are plotted in Figure 4.5. It can be seen that the magnitude and phase for the Rician distributed signal are random, but clustered around the magnitude and phase of the desired signal.

This Rician distribution is seen in all constellations. An example of QPSK is plotted in Figure 4.6. Notice that the noise signal is centered around the constellation points. That's because the signal pushes the noise to those different parts of the complex plane.

The combined signal model is illustrated in Figure 4.7. The signal has an instantaneous magnitude and phase, which provide a polar coordinate vector to a specific point in the complex plane. That point is a constellation point. The rate of change of the phase is the frequency offset. The constellation point on the complex plane is surrounded by a distribution of noise. The carrier is received at a different phase than that of the local oscillator at the receiver. The

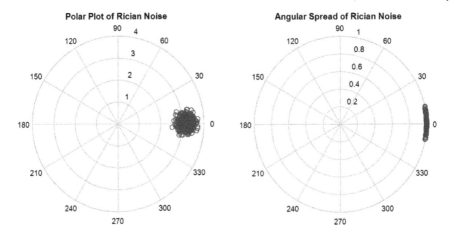

Figure 4.5 Rician Distributed Signal

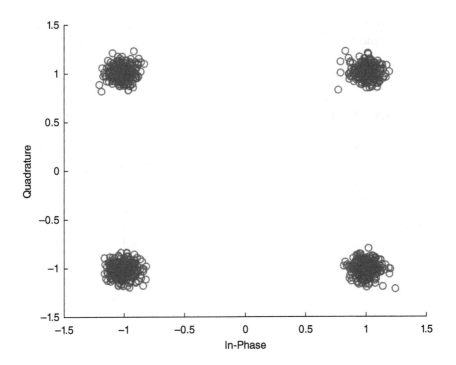

Figure 4.6 QPSK as a Rician Distributed Signal

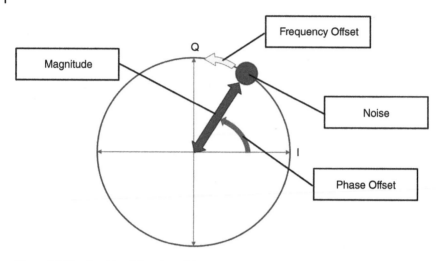

Figure 4.7 The Combined Signal Model

frequency offset and phase offset are appended to the complex-valued base-band signal model to produce equation (4.8).

$$s(t) = M(t)e^{j[\theta(t)+\omega t+\theta_0]}$$

(4.8)

4.2 Pulse Shaping

Pulse shaping is the process of turning digital data (pulses) into shapes that can be transmitted as a waveform. All symbols are derived from pulse-shaping digital data. Even a square bit is pulse shaped to be a square. Consider that the original digital-data is sampled at the data rate and has no shape, but rather pulses at logic levels. This digital data can be upsampled to a train of pulses, as shown in Figure 4.8. That train of pulses is then conditioned by the pulse-shaping filter. The result is the desired shape for the symbol.

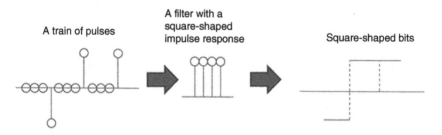

Figure 4.8 Pulse Shaping: Square Bits

A square is not a realistic shape for a symbol. This is because a square bit has infinite bandwidth. If one does not appropriately band-limit the transmitted signal, then the transmitted signal may interfere with adjacent channels. The interference with adjacent channels is called "Adjacent Channel Interference" (ACI). Wireless standards set clear limits on this unwanted out-of-channel interference. Attempting to transmit square-shaped pulses will very easily exceed limits on ACI. It is also important to note that the wireless channel is band-limited. The band-limited wireless channel may impose undesirable effects on a signal that is not appropriately band-limited. The transmit hardware will impose band-limiting. In some designs, allowing the transmitter hardware to impose band-limiting may save on costs, so long as that filtering does not impart undesirable effects.

IoT standards generally require some form of pulse shaping. This is because the best solution to band-limiting the symbol is to limit the bandwidth of the symbol before modulation. Standards for IoT protocols require different types of pulse-shaping filters. These include raised cosine, root-raised cosine, Gaussian, and half-sine. The first three types will be explored in this section. The half-sine pulse will be discussed later as it is a shift in the applied modulation technique.

Pulse shaping requires that multiple samples be used for each symbol; otherwise the symbol cannot take shape in time. Therefore, the oversampling rate is a key parameter in the design of a pulse-shaping filter. Figure 4.8 shows a symbol oversampled by four. The example in Figure 4.8 spanned only one symbol length and therefore could not smear symbols onto one another. If the filter was even one sample period longer, it would cause energy from one symbol to affect the output of another. A pulse-shaping filter that spans multiple symbols will inflict ISI.

Figure 4.8 shows interpolation being used to convert digital data into a waveform. Unfortunately, such interpolation filters spanning only one symbol will not be able to sufficiently band-limit the data. In order to effectively band-limit the data, a pulse-shaping filter will span multiple symbols.

There is a method to design a pulse-shaping filter that can prevent ISI. The Nyquist ISI Criterion provides a rule that can be used to prevent ISI in pulse shaping. The idea is that the pulse-shaping filter can be designed to have a point in the output where only one symbol contributes. The receiver will sample the pulse-shaping filter product at the symbol rate. So long as there is one point in the convolved product that has only one contributing symbol for every symbol period, the receiver can theoretically retrieve the symbol data with no ISI. In order for this to work, the pulse-shaping filter impulse response must meet the criteria in equation (4.9), such that the coefficient of the filter at any nT_s other than the current symbol is zero.

$$h(nT_s) = \begin{cases} 1 & n = 0 \\ 0 & n = 1 \end{cases} \tag{4.9}$$

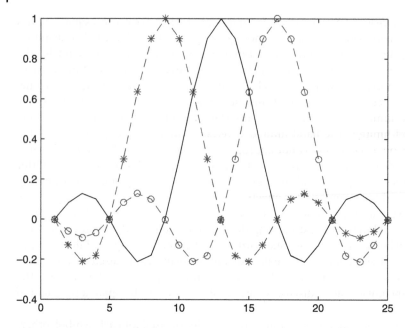

Figure 4.9 Nyquist ISI Criterion

Consider the plot in Figure 4.9. The zero crossings of sinc functions are aligned. The result is that the peak of each sinc function is aligned with contributions of zero from all other sinc functions in the plot. This illustrates the Nyquist ISI criterion.

IoT standards for linear modulation techniques generally employ either the raised cosine filter or the Root-Raised Cosine filter (RRC) as a Nyquist ISI pulse-shaping filter. Note that the convolved product of two root-raised cosine filters is a raised cosine filter.

The RRC filter can be a zero-ISI filter if convolved with another RRC filter. Standards employing an RRC filter at the transmitter therefore require an RRC filter at the receiver. This process of employing RRC filters at the transmitter and receiver allows for the use of a pulse-shaping filter that meets the Nyquist ISI criterion and also adds the benefits of "matched filtering." Matched filtering is addressed in reference [2]. Matched filters provide up to a 3 dB boost in the instantaneous SNR of a signal, as shown in [2].

Some IoT standards require a raised cosine filter at the transmitter. The raised cosine filter meets the Nyquist ISI criterion on its own. However, when convolved with another raised cosine filter, ISI is re-introduced. The use of the raised cosine filter all but precludes matched filtering. The lack of matched filtering makes the receiver less computationally expensive as compared to using the RRC filter, but foregoes the benefit of matched filtering.

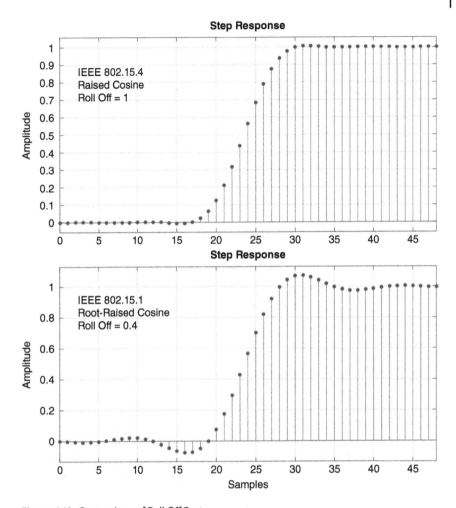

Figure 4.10 Comparison of Roll Off Factors

ISI-free pulse-shaping filters inflict overshoot and ringing in the step response of the filter. Figure 4.10 shows two pulse-shaping filters used in IoT standards for linear modulation. As the "roll off" factor increases, overshoot and ringing are reduced. This reduction on the overshoot and ringing comes at the cost of increased excess bandwidth. Figure 4.11 shows the spectrum of data output by those two pulse-shaping filters compared with the sinc-function spectrum of square-shaped pulses.

Not every wireless standard that employs pulse shaping uses ISI-free pulse-shaping filters. Some standards are designed to maximize band-limiting of

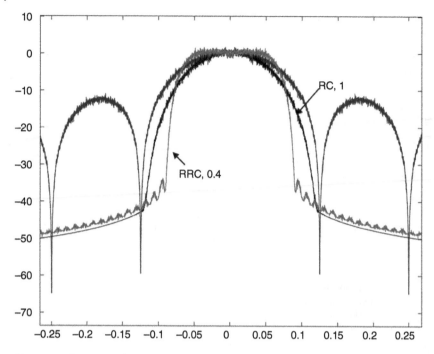

Figure 4.11 Spectrum of Pulse-Shaped Bits

the transmitted signal, foregoing the fear of ISI. This is prevalent in wireless standards that employ frequency-shift keying (FSK). The overshoot and ringing tolerated in linear modulation schemes would be problematic in FSK schemes because FSK schemes are limited by a maximum frequency deviation. A popular pulse-shaping filter for these wireless systems is the Gaussian pulse-shaping filter. The Gaussian pulse-shaping filter spans multiple symbols and does not meet the Nyquist ISI criterion. This means that the Gaussian pulse-shaping filter imposes ISI on the transmitted signal. The step response and impulse response of the Gaussian pulse-shaping filter are shown in Figure 4.12. There is no overshoot or ringing. The impulse response does not oscillate and, therefore, there is no opportunity to align zero crossings. The excess bandwidth of a Gaussian pulse-shaping filter is defined by the bandwidth time product. For Bluetooth, this value is 0.5.

The IEEE 802.15.4 standard discusses a "half-sine" pulse-shaping filter. This is actually Minimum-Shift Keying (MSK) and will be discussed later in this chapter.

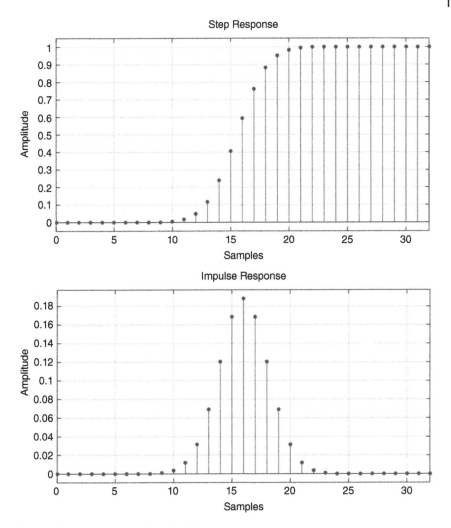

Figure 4.12 Gaussian Pulse-Shaping Filter

4.3   Modulation Techniques

There are several modulation and spread-spectrum schemes that the reader will encounter while perusing standards relevant to the Internet of Things. It will, therefore, be advantageous to provide a high-level overview of these concepts before delving deeply into the physical layer of any standard. The nomenclature and context will be reused in later sections.

Why should one modulate data? Why not transmit everything at baseband? The answer is antenna size. If one attempted to wirelessly transmit data at baseband, one would need a very large antenna. Modulation is the act of varying a parameter of a carrier signal to convey data centered at the frequency of that carrier. This has the benefit of allowing multiple access and multiplexing techniques. This also offers the ability to drastically reduce the size of the antenna used by the communication system. If the antenna is designed to be one-half the wavelength of the chosen corresponding frequency, then higher frequencies would result in smaller antennas.

Analog modulation techniques are referred to as "modulation," as in "amplitude modulation" or "frequency modulation." Digital modulation techniques are referred to as "shift keying," as in "phase-shift keying" and "frequency-shift keying." Shift keying is modulation, just done with digital data as the modulating signal. It is the digital forms of modulation that are relevant to a discussion on the Internet of Things.

In modulation schemes, the modulating signal is the signal that carries data. The carrier is the signal to be modulated. The modulating signal must embed itself by some means to the carrier signal. There are two basic types of modulation: linear and angular. This distinction comes from the representation of a signal as a complex exponential having a magnitude and a phase. The frequency of the modulated signal is the derivative of that phase with respect to time.

The modulation scheme chosen for a particular waveform defines how to convert digital data into a waveform. Digital data (bits) is compacted into "symbols" at some ratio. For binary schemes, there is one bit per symbol. Other schemes may use two or more bits per symbol. Those symbols are then mapped to the complex plane. For linear modulation, the symbols are mapped to fixed points on the complex plane. For angular modulation, the symbols may be mapped to specific motion on the complex plane. The transmitted signal does not instantaneously switch from point to point on the complex plane because that would require an infinite bandwidth. Oversampling allows the receiver to see more than one sample per symbol. This excess of samples allows the motion of the signal to be traced along paths connecting the symbols in the constellation. This will become important when synchronization recovery mechanisms are discussed. After the receiver downconverts the signal, and performs all necessary synchronization, the sample rate can be set to one sample per symbol, which will give the illusion of instantaneous transitions from symbol point to symbol point in the constellation.

Both linear and angular modulation schemes are used in the IoT. The modulation scheme chosen for a given IoT application is selected to meet the unique needs of that application. This section will explore those techniques and discuss the trade-offs incurred by using those techniques.

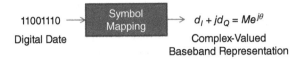

Figure 4.13 Symbol Mapping

4.3.1 Linear Modulation

Linear modulation occurs when the modulating signal is multiplied on to the amplitude of the carrier signal. The resulting product is the modulated signal and the data is primarily embedded in the amplitude of the modulated signal.

All modulation schemes in this section shall be analyzed as complex-valued baseband. The up-conversion of these modulated signals to higher frequencies is handled by transceiver architectures.

It is important to note that the data signal in linear modulation can be a complex value. The digital data is collected and mapped to different points on the complex plane (the symbol constellation), as shown in Figure 4.13. This is not the case for angular modulation where the data signal must be restricted to a real value. The fact that the data signal in linear modulation may be complex valued means that the data can modulate the carrier in quadrature. For real valued signals, this is represented with a "quadrature modulator," as shown in Figure 4.14 where the sin and cos components carry the I and Q components of the data signal. Figure 4.14 has the same structure as the Digital Up-Converter (DUC), with the exception of the sign of sine arm, which will be discussed. Therefore, a digital up-converter can be used as a quadrature modulator for a linear modulation system. Combining the symbol mapping in Figure 4.13 and the up-conversion of Figure 4.14 produces a linear modulator for any linear

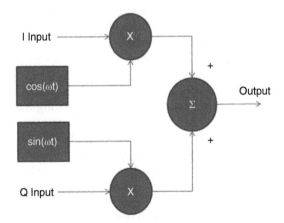

Figure 4.14 Quadrature Modulator

modulation system. Pulse shaping occurs after symbol mapping. Pulse shaping is applied to the real and imaginary values separately.

Affecting the amplitude of the modulated signal may also affect the phase of the modulated signal. When a signed real-valued or complex-valued modulating signal is used, the phase of the modulated signal may be changed with respect to the phase of the original carrier. Because of this ancillary effect on phase, the term "phase-shift keying" is often used. Unfortunately, the term "phase-shift keying" is often a misnomer because the term sometimes leads people to believe that the phase (angle) is directly being modulated. Phase-shift keying is a form of linear modulation, as will be shown in this section.

Because linear modulation schemes modulate the amplitude of a carrier, the envelope of the modulated signal varies. This can cause a wide peak-to-average power ratio and thus make amplification difficult. Linear modulation schemes generally require the use of linear amplifiers. Linear amplifiers are inherently inefficient and thus restrict battery life if used in applications for battery powered handheld devices.

Linear modulation schemes can provide better bit error rates than angular modulation. This fact means that linear modulation lends itself to higher modulation orders (more bits per symbol) compared to angular modulation.

4.3.1.1 Linear Modulation along the Real Axis

The first type of linear modulation that will be explored is real-valued. Real-valued linear modulation is when the symbol points appear along the real axis of the complex plane. There are numerous modulation schemes that place the constellation symbols solely along the real axis. Examples relevant to the IoT include Binary Phase-Shift Keying (BPSK) and M-Ary Amplitude-Shift Keying (M-Ary ASK).

BPSK places two symbols along the real axis equally distant from the origin. The constellation is shown in Figure 4.15. This configuration is called "antipodal." Reference [3] provides an excellent discussion on antipodal and orthogonal signaling. The signal moves between these points along the real axis, crossing the origin, as it switches between logic levels. This means that the signal's magnitude reduces to zero as it changes logic levels.

The signal "phase inverts" the amplitude of the carrier when it crosses the origin. This process is illustrated in Figure 4.16. By way of manipulating the sign of the amplitude of the carrier, the BPSK signal changes the phase of the carrier by 180 degrees. Figure 4.16 illustrates this phenomenon for both unfiltered and pulse-shaped cases. In the unfiltered case, the sign of the modulated signal instantaneously shifts from positive to negative. In the pulse-shaped or filtered version, the signal decreases in amplitude until it crosses the origin of the complex plane, at which time the sign switches. If noise were added, then while the signal was decreasing, there would be random shifts in sign while the aggregate signal was near the origin.

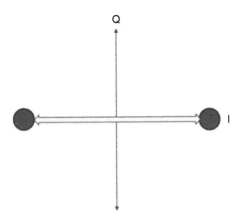

Figure 4.15 BPSK Constellation

The spectrum of a BPSK signal with square pulses is a sinc function, where the null-to-null bandwidth of the main lobe is equal to twice the data rate. The use of pulse-shaping can reduce this bandwidth. The exact amount of reduction depends on the excess bandwidth of that pulse-shaping filter.

BPSK must be demodulated coherently, meaning that the receiver must synchronize to the frequency and phase of the transmitter. An issue occurs in BPSK modulation where there is a potential 180 degree ambiguity in this synchronization. The receiver can align the BPSK symbols to the real-axis, but without additional information, it is unknown whether the first symbol is a +1 or a −1. Differential encoding provides a solution to this issue. The differential encoder is a first order feedback loop consisting of a delay and an XOR operation. The differential encoder is applied at the transmitter. The differential decoder is a feedforward XOR operation where the current bit is XOR'ed with the previous bit. The flowgraph of each is shown in Figure 4.17. The differential decoder

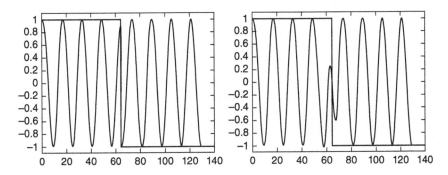

Figure 4.16 Phase Inversion of the Carrier using BPSK

XOR

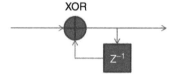

Encoder, at the transmitter

Figure 4.17 Flowgraph for Differential Encoding and Decoding

XOR

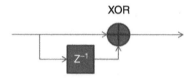

Decoder, at the receiver

is applied at the receiver. Each of these logic circuits creates a Boolean algebraic relation and each is the inverse of the other, as shown in equations (4.10) and (4.11). The symbols, s(k) are encoded from the data, d(k). The data is then decoded from the symbols. The initial symbol, s(0), is a constant.

$$s(k) = s(k-1) \oplus d(k); s(0) = C \tag{4.10}$$

$$d(k) = s(k) \oplus s(k-1) \tag{4.11}$$

Because each decoded bit decision out of the differential decoder depends on the two preceding coded bit decisions, the BER performance of Differentially encoded BPSK (DBPSK) suffers a performance hit. A single coded bit-decision error results in two decoded bit-decision errors. It is important that the receiver implement the feedforward portion. If the feedback portion were implemented in the receiver, then a single coded bit error would invert every decoded bit thereafter.

The other form of linear modulation along the real axis is M-Ary ASK. "M-Ary" means there are "m" symbol points. In M-Ary ASK, a number of $\log_2 M$ serial bits are parallelized into a word. That word is then mapped to a point on the real axis. The points on the real axis are both positive and negative and chosen to be equidistant from the nearest neighbors. For example, a 4-ary ASK modulation scheme would utilize two bits per word and map to the set of values $\{-3, -1, 1, 3\}$. This set of values can be considered a "textbook" example common in many texts that discuss ASK modulation. To delve a bit further into this type of modulation, the set of values would be "normalized" to prevent any difficulties with transmit amplifiers, making the set $\{-1, -1/3, 1/3, 1\}$. As the set of possible values gets larger, the distance between each value gets smaller. This means that in a channel with noise, the values from a large set will become blurred together even with a high SNR. This phenomenon relates to the channel capacity limit, which can be read in more detail in [3]. The set of values need not be both positive and negative. An example of this is On-Off Keying (OOK), which uses the values {0,1} and does not require coherent demodulation. However, OOK is not relevant to the IoT waveforms being discussed herein.

IEEE 802.15.4 uses both DBPSK and ASK. ASK is used as an optional modulation scheme in the IEEE 802.15.4 standard; however, the IEEE 802.15.4 standard does not follow the traditional example of M-Ary ASK as explained above. The IEEE 802.15.4 standard combines amplitude modulation with a spread-spectrum technique called Parallel Sequence Spread Spectrum (PSSS), which will be discussed in more detail later in this chapter. The important distinction is that the IEEE 802.15.4 standard does not form a word out of parallelized bits, but rather applies the sign of individual bits to individual spreading codes and then sums the output of each of those operations. The IEEE 802.15.4 ASK modulation, therefore, does not transmit individual M-Ary ASK symbols in the traditional sense. This unorthodoxy allows IEEE 802.15.4 to use large 5-bit and 20-bit words for modulation.

4.3.1.2   Quadrature Linear Modulation

Linear modulation that utilizes both real and imaginary axes of the complex plane is called "Quadrature"; 4-Ary Quadrature Amplitude Modulation (QAM) is called Quadrature Phase-Shift Keying (QPSK). QPSK can be seen as two BPSK signals that are phase orthogonal. Consider the quadrature modulator in Figure 4.14. Both the cos and sin modulated arms can be used to generate a BPSK signal. If a stream of bits were parallelized into 2-bit words, then two bits could be modulated concurrently on two phase-orthogonal carriers at the same frequency. The resulting constellation is shown in Figure 4.18. The two phase-orthogonal BPSK signals move through the origin just like standard

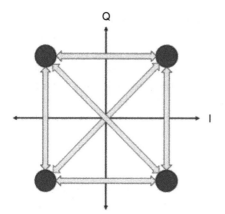

Figure 4.18 QPSK Constellation

Table 4.1 Example Mapping: Bits-to-Phase Change

Bits	Phase Change $\theta(k)$
00	0
01	$\pi/2$
10	$-\pi/2$
11	π

BPSK. Since each is independent of the other, the bit value being transmitted can change on one but not on the other, causing a shift of 90 degrees as opposed to 180 degrees. Because each symbol is based upon 2-bit words, the symbol period is twice the bit period. That is to say a new symbol is transmitted each time the transmitter is given two new bits. QPSK uses the same amount of bandwidth as BPSK, making QPSK twice as spectrally efficient.

Just like BPSK, QPSK must be demodulated coherently. QPSK suffers from phase ambiguity, as does BPSK. Differential encoding can be used to solve the problem of phase ambiguity in both BPSK and QPSK. The Boolean logic approach to differential coding employed in BPSK is insufficient for QPSK because it cannot resolve between 90 degree ambiguities. Differential coding for QPSK, therefore, embeds the bits in phase differences. This is to say that the transition between phase states conveys the information. The phase of each symbol is the sum of the phase of the QPSK mapping of the current symbol plus the QPSK mapping of every previous symbol. This process is expressed in Table 4.1 and equation (4.12). Table 4.1 shows that each symbol is mapped from a 2-bit word. The result is a phase-change value from the set $\{0, \pi/2, \pi, 3\pi/2\}$. The values presented in the table are hypothetical. The next stage in DQPSK is to sum this current phase with all previous phases, modulo 2π. This can be done by multiplying the complex values, as shown in equation (4.12). The initial symbol phase, s(0), is a constant. Note that equation (4.12) represents a feedback relationship, where the next output depends upon the previous output. The process to undo this differential encoding is, therefore, feedforward. The phase of each decoded symbol can be determined by the phase of the current coded symbol minus the phase of the previous coded symbol. For example: If the phase of the received signal is unchanged, then all logic lows have been transmitted. If the phase of the data rotates, then the direction of the rotation determines which of the two bits is logic-high. If the phase inverts, then all logic-highs have been transmitted.

$$s(k) = s(k-1)e^{j\theta(k)}; s(0) = e^{j\phi} \tag{4.12}$$

QPSK introduces a new type of ambiguity called conjugate ambiguity. This is illustrated in Figure 4.19. Matching Digital Down Converters (DDCs) and

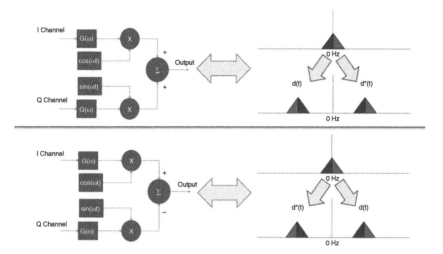

Figure 4.19 Conjugate Ambiguity

DUCs are shown in Figure 4.20. The matching is required to avoid conjugate ambiguity. The ambiguity is the result of the sign used in the summation of the quadrature modulator. A traditional DUC design, as shown in Chapter 3, pushes the conjugate image to negative frequencies. A quadrature modulator, as shown in Figure 4.14, pushes the conjugate image into positive frequencies. This difference results in constellations that are a mirror images of one another with respect to the imaginary axis. Therefore, when demodulating, it is important to know the precise mapping of 2-bit words to quadrature symbols.

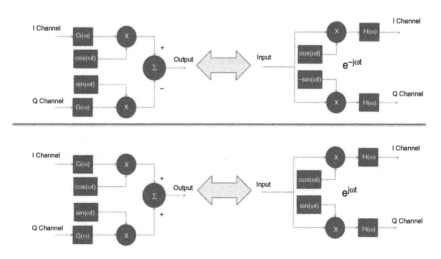

Figure 4.20 Matching DUCs and DDCs

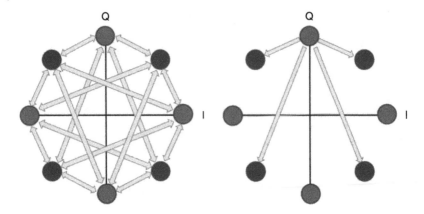

Figure 4.21 $\pi/4$-QPSK

Like BPSK, QPSK signals cross through the origin of the complex plane. Motion across constellation points through the origin results in a highly variable signal envelope. Because the signal envelope is highly variable, a linear amplifier is required at the transmitter. Linear amplifiers are far less efficient than non-linear amplifiers, and this represents a significant constraint. There are several mechanisms to mitigate the variability in the signal envelope that help to alleviate this constraint. The resulting modulation schemes are called "quasi-constant envelope." This is because these modulation schemes reduce variation in the envelope of the modulated signal. One example of such a modification to QPSK is $\pi/4$-QPSK. $\pi/4$-QPSK is called such because the entire constellation is rotated by 45 degrees ($\pi/4$ radians) after each 2-bit symbol. The resulting constellation has eight points but only four are used at any one time. The $\pi/4$-QPSK constellation is illustrated in Figure 4.21. The motion from the top-most symbol is also illustrated in Figure 4.21. Any 2-bit word is a possible next symbol, but none of the allowable transitions pass through the origin. Because the constellation has rotated, the signal will not pass through the origin regardless of where the next constellation point is.

Another means to alleviate this constraint is called offset QPSK. The two streams of bits can be offset by one-half a bit period in time. This offset in time results in a modulation scheme called offset QPSK or OQPSK. OQPSK has the same constellation as QPSK, but movement through the origin is impossible because the sign of one bit value is maintained at all times for at least half a bit period. A block diagram of the OQPSK modulator is shown in Figure 4.22. A major change between the OQPSK modulator and the QPSK modulator is the delay of $\frac{1}{2}$ a symbol period. The OQPSK constellation is shown in Figure 4.23. Notice that the symbols can no longer transition through the origin.

OQPSK is a linear modulation scheme; however, it is important to note that this term is used differently for IEEE 802.15.4. OQPSK can be modified to create

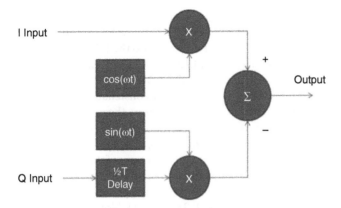

Figure 4.22 Block Diagram for the OQPSK Modulator

an angular form of modulation called MSK [4]. The term "OQPSK" may be used as a label for IEEE 802.15.4 PHY in certain bands, but in reality that modulation scheme is a type of MSK, as shall be shown.

Bluetooth uses two modulation mechanisms for Enhanced Data Rate (EDR). Those are $\pi/4$-DQPSK and 8DPSK. The "D" in DQPSK indicates that the modulation is differential. $\pi/4$-DQPSK, like all QPSK, encodes two bits per symbol; therefore, there are only four legitimate symbols for any symbol period. $\pi/4$-DQPSK is similar to $\pi/4$-QPSK and creates the same 8-point constellation from a 4-point constellation by way of rotation. The rotation is achieved by way of specific phase-changes values used for differential encoding. $\pi/4$-DQPSK uses

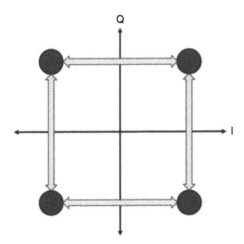

Figure 4.23 OQPSK Constellation

phase difference values from the set $\{\pi/4, 3\pi/4, -3\pi/4, -\pi/4\}$. This means that the constellation will rotate by at least $\pi/4$ (45 degrees) at every symbol period. This is how $\pi/4$-DQPSK introduces the $\pi/4$ rotation into the DQPSK modulation scheme. $\pi/4$-DQPSK follows equation (4.12), as does DQPSK.

The other Bluetooth EDR modulation method is 8DPSK. 8DPSK is differential 8PSK. As the name implies, 8PSK has an 8-point constellation. 8PSK uses 3 bits per symbol, where QPSK used only two. Therefore, there are eight valid symbol values on the complex plane for any one symbol period. For 8PSK, 3-bit words are mapped to a symbol value from the set $\{0, \pi/4, \pi/2, 3\pi/4, \pi, 5\pi/4, 3\pi/2, 7\pi/4\}$. 8DPSK achieves differential encoding by way of equation (4.12), where each phase of each symbol is the sum of the current mapping and all previous symbol phases. For 8DPSK, 3-bit words are mapped to phase-change values from the set $\{0, \pi/4, \pi/2, 3\pi/4, \pi, 5\pi/4, 3\pi/2, 7\pi/4\}$. Notice that a phase change of 0 is possible. This means that the constellation is not rotated. Figure 4.24 shows the constellation of both 8PSK and 8DPSK. Neither 8PSK nor 8DPSK rotates the constellation. Each symbol value can transition to any other symbol value. Therefore, both modulation schemes allow transitions through the origin.

4.3.2 Angular Modulation

Angular modulation occurs when the modulating signal is applied to the angle, or phase, of the carrier signal. Angular modulation is constant-envelope, meaning that the magnitude of the carrier is not affected by the modulating

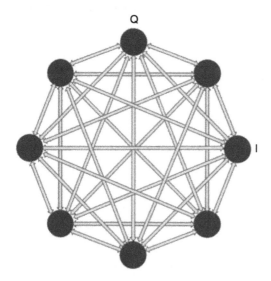

Figure 4.24 8PSK and 8DPSK Constellation

signal. Angular modulation, by its nature of affecting the angle of the carrier, requires non-linear effects in the modulator.

If the modulating signal modulates phase, then the signal can be expressed as in equation (4.13).

$$s(t) = Me^{i[\omega t + \Delta \phi d(t)]}$$ (4.13)

If the modulating signal modulates frequency, then the signal can be expressed as in equation (4.14).

$$s(t) = Me^{i[\omega t + \Delta f \int d(t) dt]}$$ (4.14)

As was shown in the case of BPSK, "phase-shift keying" occurs when linear modulation crosses the origin of the IQ plane. Frequency modulation is specifically the modulation of the frequency of a carrier signal. Phase modulation is specifically the modulation of the phase of a carrier signal. Phase modulation is a rarity. This is because frequency modulation provides superior benefits. Phase is constrained to modulo 2π. Frequency has no such constraint.

Non-linear amplifiers can be used in transmitters for systems employing angular modulation because angular modulation is constant-envelope. Non-linear amplifiers are more power efficient than linear amplifiers, and this fact makes non-linear amplifiers attractive for power-constrained and battery-operated systems.

4.3.2.1 Frequency-Shift Keying

The basic concept of all frequency modulation is illustrated in Figure 4.25. This concept is to convert values of amplitude into values of frequency. The amplitude of the modulating signal is constrained between −1 and + 1. That amplitude is then multiplied by the "frequency deviation" Δf. This product then drives the center frequency of the modulated signal. The effect is shown in Figure 4.25 where the modulating signal is plotted over the modulated signal. When the modulating signal has high amplitude, the modulated signal has high

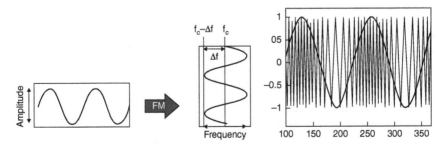

Figure 4.25 Frequency Modulation

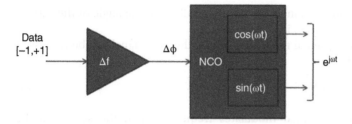

Figure 4.26 Continuous Phase FSK

frequency. When the modulating signal has low amplitude, the modulated signal has a low frequency.

For Frequency-Shift Keying (FSK), the modulating signal has discrete states and thus discrete frequencies are transmitted. For binary systems, these frequencies (or "tones") are sometimes referred to as the "mark" and "space" frequencies, which are terms harkening back to the days of the telegraph.

There are numerous ways to generate a frequency modulated signal. FSK can be generated by driving an analog Voltage Controlled Oscillator (VCO) in hardware, or by driving a digital Numerically Controlled Oscillator (NCO) in software or firmware. This technique is called Continuous Phase FSK (CPFSK). The CPFSK process is shown in Figure 4.26. The modulating signal, with amplitude limited to $[-1, +1]$ is conditioned by multiplying the amplitude by the frequency deviation. That conditioned modulating signal is then used to drive an NCO. The signal is referred to as the "phase increment" $\Delta\phi$. Phase increment is the frequency at which the NCO is driven. The NCO accumulates the phase increment over time and thus approximates the integration over time. The modulated signal can be created as a complex-valued baseband signal and then upconverted for channelization through a DUC. DUCs were discussed in the Radio chapter.

The number of discrete amplitudes supplied to the modulator determines the number of FSK "tones," or distinct frequencies that will be transmitted. A binary FSK system supplies $+1$ and -1. A 4-ary FSK system would supply $\{-1, -1/3, +1/3, +1\}$.

The NCO is sometimes referred to as a Direct-Digital Synthesizer (DDS). DDS is discussed in detail in [5]. A detailed discussion on Continuous Phase Modulation (CPM) is beyond the scope of this book. An excellent reference on CPM techniques is provided in the IEEE papers [6], [7], and also [8].

4.3.2.2 Modulation Index and Bandwidth

An important parameter for frequency modulation is the modulation index, h. The calculation of h for digital systems is given by equation (4.15). The variable R is the symbol rate and the variable Δf is one-half the spacing between tones.

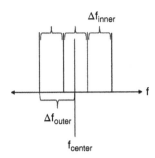

Figure 4.27 Inner and Outer Frequency Deviation

One-half of the spacing between tones is the frequency deviation, or "inner" frequency deviation in the case of FSK modulation with more than two tones. The "inner" frequency deviation is one-half of the spacing between tones and it assumes the tones are evenly spaced. The outer frequency deviation is one-half of the spacing from the center frequency to the farthest FSK tone. This concept is illustrated with the four tones from a 4-ary FSK system in Figure 4.27. For a binary FSK system, there is only the inner frequency deviation and the data rate is the symbol rate. Therefore, the calculation of the modulation index simplifies to equation (4.16).

$$h = \frac{2\Delta f_{inner}}{R_{Sym}} \tag{4.15}$$

$$h = \frac{2\Delta f}{R} \tag{4.16}$$

The modulation index determines whether the frequency modulation is "wide-band" or "narrow-band." A wide-band frequency modulated signal, such as broadcast FM, has a large modulation index. A large modulation index results in a bandwidth much larger (severalfold) than the information bandwidth but has benefits at the receiver. "Narrow-band" frequency modulation has a modulation index less than unity.

The modulation indices for Bluetooth are shown in Table 4.2. Bluetooth Low Energy (LE) specifies a modulation index between 0.45 and 0.55. Bluetooth "Basic Rate" requires a modulation index between 0.28 and 0.35. Both the LE

Table 4.2 Bluetooth Modulation Indices

Standard	Frequency Deviation	Modulation Index
Basic Rate	157.5 kHz	$0.315 \pm 11\%$
Low Energy	250 kHz	$0.5 \pm 10\%$

Table 4.3 ITU-T G.9959 Modulation Indices

Data Rate	Frequency Deviation	Modulation Index
19.2 kbps (R1)	20 kHz ± 10%	1.0415
40 kbps (R2)	20 kHz ± 10%	1
100 kbps (R3)	29 kHz ± 10%	0.58

and basic rate variants have a data rate of 1 Mbps. Both use Gaussian pulse shaping.

ITU-T G.9959 uses three different data rates and specifies a frequency deviation for each, as shown in Table 4.3. The FSK schemes for R1 and R2 rates do not use any baseband filtering. The data for the R1 rate is Manchester encoded. Manchester encoding puts a level change in every bit. Manchester encoding helps with symbol timing recovery and prevents a DC offset that would be caused by long strings of the same value. The level change in every bit means that the baud rate is twice the bit rate, and the baud rate will be used to approximate the modulation index for R1 in Table 4.3. The FSK scheme for R3 uses Gaussian pulse shaping, which will be discussed shortly.

It should be noted that a modulation index of 0.5 produces a very special type of FSK called MSK. The definition of OQPSK in IEEE 802.15.4 follows the paradigm for minimum-shift keying, which will be discussed later.

Determining the bandwidth of a frequency modulated signal relies on the data rate and the modulation index. Carson's Rule, shown in equation (4.17), provides a means for determining the bandwidth of approximately 98% of the power in a frequency modulated signal. The variable Δf is the frequency deviation and the variable f_m is the highest frequency component in the modulating signal.

$$BW_{98\%} = 2(\Delta f + f_m) \tag{4.17}$$

For binary FSK signals, Carson's Rule is given by equation (4.18), where R is the data rate.

$$BW_{98\%} = 2\Delta f + R \tag{4.18}$$

The modulation index has an impact on the type of demodulator that is to be used. A modulation index of 0.5 represents the minimum tone spacing required to maintain orthogonality between the tones in a coherent correlator receiver [3]. A modulation index of 1 is required to maintain orthogonality between the tones in a non-coherent correlator receiver. If orthogonality between the tones is not maintained, then energy from one tone will be partially detected in

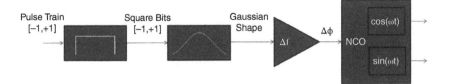

Figure 4.28 GFSK Modulation

the correlator of another tone. This leakage will worsen bit error performance, potentially significantly.

4.3.2.3   Gaussian Pulse Shaping
Pulse shaping for angular modulated signals must be performed before NCO. NCO is a non-linear operation on the modulating signal, where the modulating signal is interpreted as the angle of the NCO output. The Gaussian pulse has no overshoot in the step response, and therefore lends itself to FSK systems for which the bandwidth would be sensitive to the extra bandwidth caused by ripples in the amplitude of the filtered modulating signal. FSK that employs a Gaussian pulse-shaping filter is called Gaussian FSK (GFSK).

A modulator using CPFSK for GFSK is shown in Figure 4.28. Note that the pulse train representing the data to be transmitted must first be converted to square bits. This initial rectangular pulse shaping is done with a moving average filter. The Gaussian filter does not shape "pulses" but rather rectangular logic levels. To this end, some implementations will convolve the Gaussian filter with the moving average filter rather than employing two filters, as shown in Figure 4.28.

Gaussian pulse shaping helps restrict the bandwidth of the FSK signal significantly. Bluetooth and data rate R3 in ITU-T G.9959 use Gaussian filtering to constrain the bandwidth of the modulated signal. By contrast, the relatively low data rates ITU-T G.9959 R1 and R2 use unfiltered FSK. This causes those two modes to spread energy beyond the date rate of the original modulating signal; however, given the low data rates of those two modes (R1 and R2), bandwidth constraints are not necessary.

The Gaussian filter does not meet the Nyquist ISI criterion. This filter imposes inter-symbol interference. The effect of this sub-optimality on bit error performance can be mitigated by way of more complex algorithms at the receiver.

4.3.2.4   Minimum-Shift Keying
When the modulation index of an FSK signal is set to 0.5, this is called "Minimum-Shift Keying." This is the minimum modulation index to maintain orthogonality on a coherent FSK correlation receiver [3].

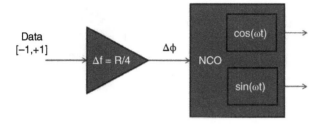

Figure 4.29 Differential MSK by CPFSK

Reference [4] gives an excellent review of MSK as a modulation scheme. Reference [4] uses a modified OQPSK modulator to produce MSK. There are also many other implementations of MSK. For example, the CPFSK modulator from Figure 4.26 can also meet the MSK criterion of a modulation index of 0.5. Bluetooth LE uses an MSK method that matches Figure 4.26. It should be noted that MSK waveforms produced by the CPFSK modulator and reference [4] do not match. However, both are MSK waveforms. Furthermore, neither of the aforementioned two MSK methods matches the waveform output by the method used by IEEE 802.15.4 called "OQPSK." IEEE 802.15.4 defines two modes for "OQPSK." One of those two modes is an OQPSK modulator modified with half-sine pulse-shaping filters. This type of pulse shaping produces an MSK waveform. Despite the fact that all methods result in a modulation index of 0.5, these waveforms are not interchangeable without some data manipulation. To constrain this discussion to modulation schemes specific to IoT, only the IEEE 802.15.4 and Bluetooth LE forms of MSK will be detailed.

MSK can be created by way of a CPFSK modulator. That process is shown in Figure 4.29. The frequency deviation is set to a value equal to one-quarter of the data rate and the modulator is run as described before. This results in a binary FSK signal that meets the minimum-shift keying criterion. This method is called "Differential MSK" because when demodulated by the aforementioned modified OQPSK demodulator, the resulting output bits are differentially encoded. If demodulated by a frequency discriminator, the resulting output bits are not differentially encoded. The modulator in Figure 4.29 is the modulator used for Bluetooth LE. Using this differential form of MSK allows the waveform to be demodulated non-coherently, such as through a frequency discriminator.

The output of this MSK modulator is shown in Figure 4.30. The modulating signal is encoded in the instantaneous frequency of the modulated signal. Notice that the phase transitions only occur at half-period points. This resembles an OQPSK scheme with half-sine shaped pulses.

The IEEE 802.15.4 physical layer standard specifies an OQPSK scheme that results in a different type of MSK than those obtained through differential or

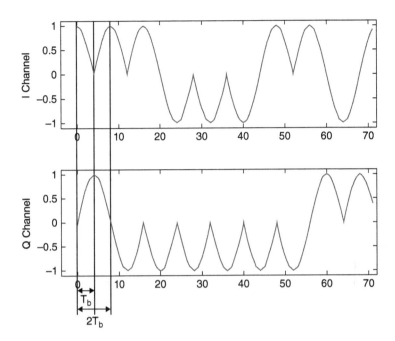

Figure 4.30 CPFSK MSK Modulator Output

non-differential techniques. This MSK modulator is illustrated in Figure 4.31. The symbol for this modulator takes two bits; therefore, the symbol period is two times the bit period. The I and Q bits are pulse shaped with a half-wave sinusoidal pulse-shaping filter. The bit pulses take two times the bit period to shift through the sinusoidal filter. The Q-arm is delayed by one-half a symbol, which is equal to one bit period. For 802.15.4, the bit period applied at this stage

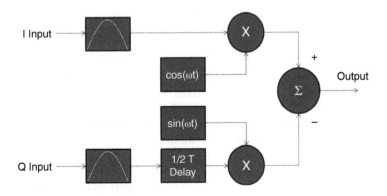

Figure 4.31 802.15.4 OQPSK

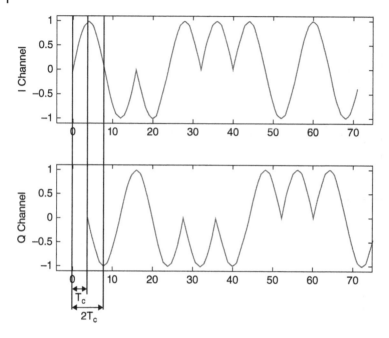

Figure 4.32 IEEE 802.15.4 OQPSK Output

is equal to the "chip rate." OQPSK in 802.15.4 uses direct-sequence spread spectrum, which will be discussed in Section 4.5.2. The complex-valued baseband output of this modulator is shown in Figure 4.32. The Q-arm output is delayed by one chip. The half wave sinusoid shape spans two chip periods.

4.3.3 Probability of Symbol Errors

One of the most important factors to consider when choosing a physical layer is the probability of symbol error. The probability of symbol error determines the error rate. The error rate curtails throughput. This phenomenon requires higher layer functions to mitigate the effect and such higher layer mitigation strategies will be discussed at greater length in the Media Access chapter.

The probability of symbol error is specific to the receiver, not the modulation. The choice of modulation type does allow or disallow certain types of receivers, but it is ultimately the receiver, and more specifically the implementation of that receiver, that determines the probability of symbol error.

Symbol errors happen when the demodulator incorrectly determines the value of a received symbol. There are numerous types of distortions affecting the ability of the receiver to measure the value of the signal and demodulate that signal to data. Additive White Gaussian Noise (AWGN) at the receiver affects the ability of the demodulator to distinguish between different symbols. Small-scale fading can cause sudden drops in signal power, and when coupled with AWGN, can further frustrate the ability of the demodulator to distinguish between different symbols. Inter-symbol interference, either caused by a multipath channel or caused by the modulation scheme can smear symbols together, making symbol disambiguation difficult. Phase and frequency offsets make the received signal constellation rotate in the complex plane, and when coupled with static decision regions, cause the demodulator to misidentify symbols. Non-linear distortion through the receiver, such as clipping at the analog-to-digital converter, causes unwanted spectral components, which cause ambiguity among received symbols. Interference in a congested band can cause disruptions in the received signal, causing more symbol errors. It is, therefore, important that when specifying a probability of bit error, one also specify the model of the receiver, imperfections thereof, and the type of channel used.

Constant envelope modulation is advantageous for the transmitter because it allows the use of non-linear amplifiers. This is not as important for the receiver. This section will focus on symbol errors at the receiver.

4.3.4   Correlation Receivers

Correlation receivers are covered in detail in references [3] and [9], among other notable textbooks on the subject of wireless digital communications. The structure is the same for linear and nonlinear modulation. The incoming signal is correlated with a known waveform. The result of the correlation then provides a measurement of symbol magnitude and phase.

Consider the correlation demodulator illustrated in Figure 4.33. There are two correlators, each dedicated to the same frequency but 180 degrees out of phase. The benefit is maximized for antipodal signaling in BPSK. When the BPSK signal is at 0 degrees, the logic-high correlator returns a positive value and the logic-low correlator returns a negative value. When the BPSK signal is at 180 degrees, the opposite is true. For correlation demodulator frequency-shift

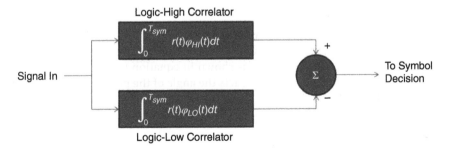

Figure 4.33 Correlation Demodulator

keying represents orthogonal signaling. When a BFSK signal transmits a logic-high, the logic-high correlator returns a positive value but the logic-low signal returns nothing but noise. This is because it is assumed that the two frequencies chosen to represent the logic levels in the BFSK signal are orthogonal.

BPSK, as an antipodal signal, enjoys twice the energy per bit than BFSK through the correlation demodulator. This is because both logic-high and logic-low correlators successfully correlate with the BPSK signal, just with opposite polarity. The BFSK signal can only have one correlator successfully correlating with the signal while the other correlator becomes a source of noise.

The correlation demodulators assume that symbol timing has been recovered and that carrier frequency has been recovered. Receiver synchronization will be discussed in detail in Section 4.4. If symbol synchronization has not taken place, then the timing of the correlators will be in error. If carrier synchronization has not taken place, then the output of the correlators will suffer in magnitude and may become phase inverted over time.

4.3.5 Arctan Demodulator

Frequency modulation schemes such as FSK and MSK can be demodulated by taking the derivative of the angle of the modulated signal. The signal is constant envelope with a varying angle, as defined in equation (4.19).

$$s(t) = Me^{j\theta(t)} \tag{4.19}$$

The angle function extracts the angle of s(t). The variable $\theta(t)$ is the instantaneous phase of s(t). The derivative of that angle yields the instantaneous frequency, m(t), as shown in equation (4.20).

$$m(t) = \frac{d\, \text{angle}\{s(t)\}}{dt} = \frac{d\theta(t)}{dt} \tag{4.20}$$

When implemented digitally, the derivative can be represented as a first order difference, as shown in equation (4.21).

$$\Delta\theta = \theta[n] - \theta[n-1] \tag{4.21}$$

This demodulation technique can be achieved using an "arctan demodulator," as shown in Figure 4.34. The incoming signal is multiplied by the conjugate of that signal delayed by one sample, as shown in equation (4.22). The arctan of the product is then taken, which extracts the angle of the product, as shown in equation (4.23).

$$e^{j\theta[n]}e^{-j\theta[n-1]} = e^{j\Delta\theta[n]} \tag{4.22}$$

$$\text{atan}\left(\frac{\text{imag}\{e^{j\Delta\theta[n]}\}}{\text{real}\{e^{j\Delta\theta[n]}\}}\right) = \text{angle}(e^{j\Delta\theta[n]}) = \Delta\theta[n] \tag{4.23}$$

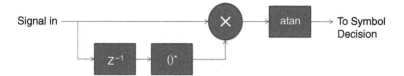

Figure 4.34 Arctan Demodulator

It is important to note that this process is nonlinear. Because the process is nonlinear, thermal noise no longer follows a Gaussian distribution. Noise will follow the distribution of the phase of a Rician distribution, as discussed in Section 4.1.2.

4.3.6 Comparison of Arctan and Correlation Receivers for FSK

The two types of distortion that will be considered here are noise and phase/frequency offset. Reference [10] provides a study into these differences. These results from reference [10] are illustrated in Figure 4.35. Part A of Figure 4.35 shows the results of the arctan receiver. The "Digital Cross-Differentiate-Multiplier" (DCDM) receiver is a modified form of the arctan receiver and will be discussed in Section 4.3.7. The data rate for the simulation in reference [10] was 1000 kbps and the modulation index was 8.

The correlation demodulator has a superior lower bound to BER performance when compared to the arctan demodulator. The BER curve of the correlation demodulator, provided there is perfect carrier and timing recovery, is optimal. The correlation demodulator requires that carrier and timing recovery take place before demodulation. Carrier and timing offsets will severely distort the output of the correlation demodulator. As is shown in Figure 4.35, the BER performance of the correlation receiver collapses in the face of carrier offsets. Carrier and timing recovery are discussed in Section 4.4.

The arctan demodulator does not require symbol timing to be resolved before demodulation. The signal can be demodulated first, and symbol timing recovered after the fact. The arctan demodulator shows resilience to carrier offsets. The instantaneous phase of the modulated signal, s(t), is defined as the accumulation of the instantaneous frequency, m(t), with an offset frequency, ω, as shown in equation (4.24).

$$s(t) = Me^{j\theta(t)} = Me^{j\{\omega t + \int_0^t m(\lambda)d\lambda\}} \tag{4.24}$$

The derivative of the angle then yields m(t) summed with a DC offset defined by ω, as shown in equation (4.25).

$$\frac{d\,\text{angle}\{s(t)\}}{dt} = m(t) + \omega \tag{4.25}$$

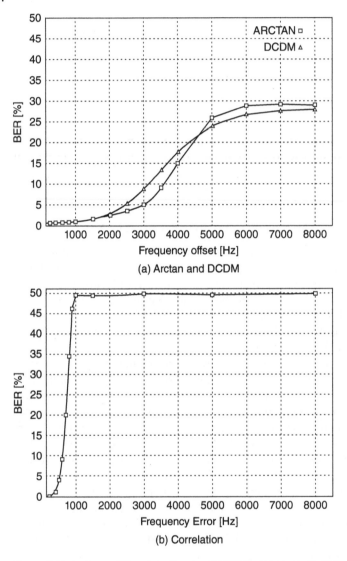

Figure 4.35 Arctan and Correlator Receiver BER Performance vs. Carrier Offset [10]

When the receiver is synchronized, meaning that carrier and timing offsets have been corrected, the correlation detector yields superior symbol error performance when compared to the arctan demodulator. However, the receiver will not always be synchronized. It may be that the low-power devices used for a wireless IoT system need to be inexpensive and that may preclude accurate carrier and timing estimates. Additionally, with a strong line-of-sight and short distance, the received SNR will be high. In such a case, the arctan demodulator might be a better choice.

4.3.7   Efficiency of the Arctan Demodulator

The arctan function is a transcendental function, meaning that it cannot be expressed in a finite sequence of algebraic operations. This leads to the question of how to implement an arctan function. There are several ways to accomplish this, including CORDIC (COordinate Rotation DIgital Computer) [11] and look-up tables.

Alternatively, one can implement the derivative of arctan [12]. This is shown in equation (4.26). Arctan itself may be transcendental, but the derivative of arctan can be represented through a finite sequence of operations. Equation (4.26) leads to what reference [10] called the "Digital Cross-Differentiate-Multiplier." Given that the arctan demodulator is based on the derivative of arctan, this design is highly advantageous.

$$\frac{d \arctan\left(\frac{Q}{I}\right)}{dt} = \frac{I(t)\dfrac{dQ(t)}{dt} - Q(t)\dfrac{dI(t)}{dt}}{I^2(t) + Q^2(t)} \tag{4.26}$$

That demodulator is illustrated in Figure 4.36. A complex-valued signal is input into the demodulator. The in-phase arm is multiplied by the time derivative of the quadrature phase arm, and the quadrature phase arm is multiplied by the time derivative of the in-phase arm. The normalization operation, in which the difference between those two arms is divided by the instantaneous magnitude of the signal, is not shown.

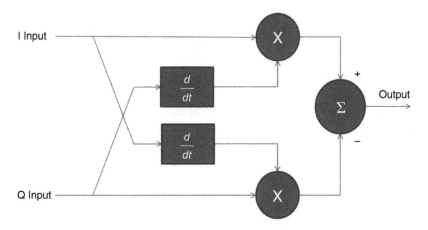

Figure 4.36 Cross-Differentiate-Multiplier FSK Demodulator

4.4 Synchronization

What is synchronization, and why is it important? Synchronization is the means by which the receiver "synchronizes" itself to the transmitter such that the received signal can be properly demodulated. This synchronization takes the form of carrier synchronization, symbol synchronization, and frame synchronization.

In order to demodulate a signal, there are several parameters that must be determined. The demodulator must be able to determine the start of the transmission. The demodulator must be able to determine the center frequency, and possibly the phase, of the carrier of the transmission. The demodulator must be able to determine the symbol timing of the transmission. All of these determinations are obtained by way of synchronization.

This section will establish the importance of synchronization in the receiver and discuss common methods to achieve that synchronization.

4.4.1 Frame Synchronization

Consider a stream of bits being received. Assuming the receiver can successfully demodulate the signal into bits, the receiver must then organize those bits into bytes. How can the receiver determine the start of a byte? The receiver cannot assume when a transmission has begun. The receiver will need some mechanism to determine the start of a frame of transmitted data. Now consider a "burst" system where the data is transmitted in separate segments with silence in between. The receiver needs to be able to determine that a burst of data has occurred, as opposed to the noise in between bursts.

This problem of determining the start of a transmission can be solved by concatenating a "synchronization word" (sync word) on to a packet of data to allow the receiver to determine the start of that packet. The idea is illustrated in Figure 4.37. Bluetooth, ITU-T G.9959, and IEEE 802.15.4 all use this paradigm. There are two synchronization words appended to the packet, a sync word and a "preamble." It is important to note that this terminology is not used consistently across wireless standards. The use of the term "preamble" used in Figure 4.37 reflects the terminology of Bluetooth, IEEE 802.15.4, and ITU-T G.9959.

The preamble, as shown in Figure 4.37, is used to train carrier recovery mechanisms like phase-lock loops. Carrier recovery will be discussed in Section 4.4.1.3. For Bluetooth and ITU-T G.9959, the preamble is an alternating sequence of ones and zeros. For most variations of IEEE 802.15.4, the preamble is a string of logic-low values.

Figure 4.37 Sync Word in Data Packet

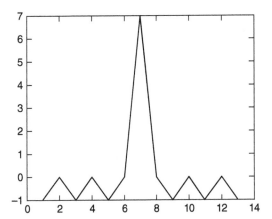

Figure 4.38 Seven-Bit Barker Code Autocorrelation

The sync word, as shown in Figure 4.37, is a sequence of values that exhibit low autocorrelation side lobes. This allows a correlator at the receiver to perform a cross-correlation and output an impulse-like function if the sync word is present. The correlation of the sync word has additional benefits besides determining the start of the transmission. One of the first treatments on the subject of determining the start of a transmission was written by Ronald Hugh Barker in 1953 [13]. The result of that work is what we today call "Barker Codes." Barker Codes are binary sequences that exhibit a singular peak and low side lobes when autocorrelated. Consider the sequence {1,1,1,−1,−1,1,−1}. The result of the autocorrelation of this sequence is shown in Figure 4.38. Notice that there is one peak and the side lobes are well below that peak. This gives the autocorrelation output an "impulse-like" shape.

This impulse-like output of the correlator has a sharp peak. This peak can be used for coarse symbol timing recovery. The more oversampled the sync word when going through the correlator, the more resolution is provided in that coarse symbol timing.

The impulse-like output can also be used to recover the wireless channel impulse response without the need of an equalizer. This concept is illustrated in Figure 4.39. The equations are worked out in equations (4.27)–(4.29). The transmitted synchronization word has been convolved with the channel impulse response; therefore, the impulse-like shape resulting from the autocorrelation of the sync word automatically yields an estimate of the impulse response of that channel.

$$b_{tx}(t) \circledast b_{rx}(t) = \delta(t) \qquad (4.27)$$

$$b_{tx}(t) \circledast h(t) = s(t) \qquad (4.28)$$

$$b_{rx}(t) \circledast s(t) = \delta(t) \circledast h(t) = h(t) \qquad (4.29)$$

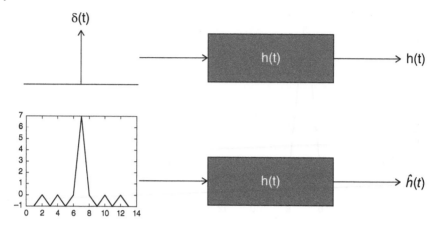

Figure 4.39 Recovering the Channel Impulse Response through Frame Synchronization

Most of the IEEE 802.15.4 modes use an 8-bit sequence {1,1,1,−1,−1,1,−1, 1}, which almost matches the 7-bit barker code. That autocorrelation output is shown in Figure 4.40. This 8-bit sync word resembles the original 7-bit barker code closely enough for synchronization purposes while allowing a full byte to be transmitted.

4.4.2 The Difference between Carrier and Symbol Synchronization: BPSK Example

This section will consider the case of synchronizing the carrier and symbol timings of a linearly modulated BPSK signal. BPSK is chosen because low order

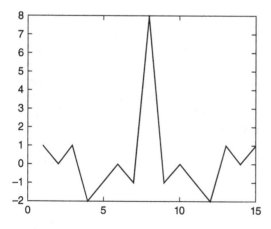

Figure 4.40 IEEE 802.15.4 Sync Word Autocorrelation

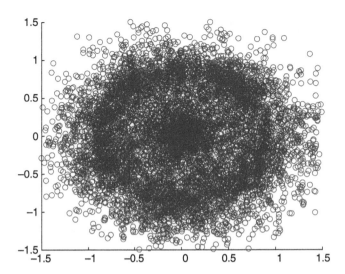

Figure 4.41 BPSK with Carrier and Symbol Timing Offset

linear modulation provides a means to easily visualize the effects of different synchronization errors. Frame synchronization will be ignored for this example.

Consider the example of a BPSK signal with noise. The received BPSK signal has an unknown carrier offset and an unknown symbol timing offset. Such a received signal is plotted on the complex plane in Figure 4.41. The signal is unrecognizable as a BPSK signal. The signal appears to be a giant ball of noise. This is because the carrier frequency offset is spinning the constellation around the origin of the complex plane and the symbol timing offset is preventing samples from being taken at optimal symbol sampling points. The signal is being sampled in transitions between the two symbols.

If the receiver implements a method to correct for the carrier offset, but not the symbol timing offset, the signal will appear as a line of noise smeared across the real-axis. This is illustrated in Figure 4.42. The rotation around the origin of the complex plane has been resolved; however, samples are still being taken in the transition between the two symbol points. The fact that the signal is smeared along the real-axis means that the receiver has synchronized to both carrier frequency and phase.

If the receiver implements a method to correct for the symbol timing offset, but not the carrier offset, the signal will appear as a ring of noise rotating around the origin of the complex plane. This is illustrated in Figure 4.43. The samples are no longer taken in the transition between the two symbol points; however, those symbol points are rotating. The fact that the signal is being sampled at

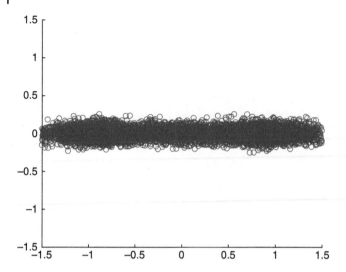

Figure 4.42 BPSK with Symbol Timing Offset, Carrier Corrected

the optimal symbol point as the constellation rotates means that the receiver has synchronized to the symbol rate and symbol phase.

When symbol timing and carrier synchronization are applied, the received BPSK signal returns to the familiar two-point constellation. This is illustrated in Figure 4.44. The two symbol points have a cluster of noise around them, as is to be expected.

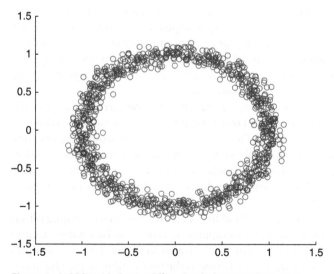

Figure 4.43 BPSK with Carrier Offset, Symbol Timing Corrected

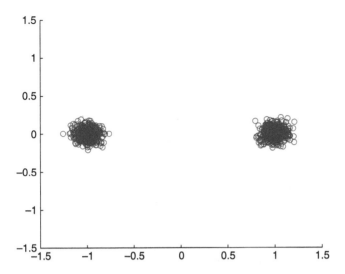

Figure 4.44 BPSK Carrier and Symbol Timing Corrected

4.4.3   Carrier Synchronization

The complex-valued baseband model of the received signal established in Section 4.1.3 allows for a "phase offset" and a "frequency offset." The phase offset exists because the phase of the carrier signal will not match that of the local oscillator of the receiver. A transmit and receive oscillator, even if at the same precise frequency, will be of a different phase. Propagation delay will also give rise to a carrier phase difference between the transmitter and receiver. An offset in phase will mean that a modulation scheme such as BPSK will not be placed along the real axis, as described in Section 4.3.1.1. The phase mismatch will cause the BPSK vector to have an angle, and unless that mismatch is corrected, the energy per symbol will be decreased when projected onto the real axis, as shown in Figure 4.45. The frequency offset exists because there will be a mismatch between the carrier frequency and the frequency of the local oscillator at the receiver. Oscillators produce a tone at a given frequency within an established tolerance. This tolerance provides a range, however small, for an error in the actual frequency produced. If the receiver or transmitter is moving with respect to the other, then the Doppler Effect will cause additional frequency errors. This means that two oscillators of the same frequency will produce tones with measurable differences in frequency. This has the effect of rotating the constellation in the complex plane. In order to align the received signal with the expected constellation, the phase and frequency offsets must be corrected.

The effect of a frequency offset accumulates in the phase offset. Frequency is the time derivative of phase; therefore, the phase offset of a signal will grow

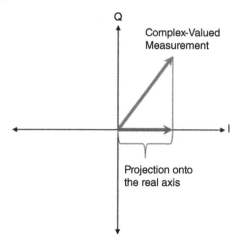

Figure 4.45 Projection onto Real-Axis due to Phase Offset

according to the frequency offset. An initial phase offset may be small, or an initial phase offset may be corrected, but the frequency offset will cause the phase offset to grow over time. Therefore, a frequency offset will cause the loss of signal energy shown in Figure 4.45 to worsen over time.

If only carrier frequency is resolved, and not carrier phase, the signal would have a slant, as shown in Figure 4.46. The constellation is slanted in the complex plane. Carrier frequency has been synchronized, so the signal is no longer rotating.

What about correcting carrier phase and not carrier frequency? A carrier frequency offset will drive the receiver and signal phase out of synchronization. A periodic update of the phase at the receiver can itself be construed as a form of carrier frequency correction, given that such a process has a non-zero time

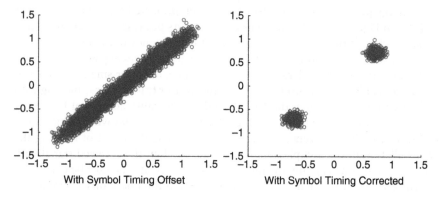

Figure 4.46 BPSK with Carrier Phase Offset

Signal

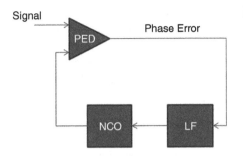

Phase Error

Figure 4.47 Basic PLL Structure

derivative. If the carrier frequency offset is small relative to the update rate with which the receiver updates its local version of the carrier phase, then such a phase recovery mechanism is also a frequency recovery mechanism. If phase offset is periodically corrected, simply by observing the error in the received constellation over time, then the receiver can implement an inexpensive form of carrier synchronization. This type of mechanism is a Type I Phase-Lock Loop (PLL). PLLs are discussed in great detail in reference [14]. The basic structure of a PLL is shown in Figure 4.47. An incoming signal is compared to an expected carrier generated by a Numerically Controlled Oscillator (NCO). The Phase Error Detector (PED) measures the error between the phases and produces the error signal. The error signal is filtered by the loop filter (LF) and that filtered error signal is used to drive the NCO.

NCOs are discussed in Section 4.3.2. The loop filter will be a low pass filter. The implementation of the loop filter can be used to determine the "type" of PLL. Using the taxonomy in [14], a Type I PLL is a PLL with one accumulator that can correct for phase offset. A Type II PLL has two accumulators and can correct for frequency offset. For the PLL shown in Figure 4.47, there is one accumulator in the NCO. A type II PLL will have an accumulator in the loop filter. There are multiple ways a phase error detector can be implemented. The implementation of the phase error detector determines the value of the error signal as a function of the phase difference between the incoming signal and the locally generated carrier.

Higher order modulation schemes can complicate the estimate of the carrier offset. The BPSK example that has been illustrated thus far offers the simplicity of having symbol information embedded in the amplitude, leaving the angle for carrier offset information. Even with that simplicity, the symbol information affects "amplitude" and not magnitude. Therefore, the PLL must be immune to 180-degree phase shifts. This type of PLL is called a Costas Loop.

As the modulation scheme becomes more complex, more symbol information becomes embedded in the angle of the received signal. The increasing complexity is beyond the scope of this book. Reference [15] has a chapter that details this increasing complexity very well.

Differential modulation schemes provide immunity to phase ambiguity, as discussed in Sections 4.3.1 and 4.3.2. If the receiver employs phase differentiation, then a differential modulation scheme also provides immunity to static carrier phase offsets. A receiver based on phase differentiation does not project the signal onto axes on the complex plane but rather examines the average of the trajectory of phase. Frequency-shift keying also provides immunity to static carrier phase offsets for this reason. Frequency-shift keying and differential phase-shift keying both employ a non-zero time derivative to the phase of the signal.

The output of the arctan demodulator, discussed in Section 4.3.5, contains a DC offset representative of the carrier frequency offset. This DC offset can be used to tune the receiver's center frequency. If the demodulated output is put through a tight lowpass filter designed to isolate the DC component, the output of that filter can be looped back to the channelization process, providing a PLL.

4.4.4 Data Whitening

Data whitening is employed in wireless IoT protocols employing frequency-shift keying, like Bluetooth and ITU-T G.9959. If a receiver employs a carrier synchronization loop such as a PLL, the PLL may respond to a long string of ones or zeros by over-correcting. This is to say that the PLL will inadvertently place the carrier frequency at the frequency of the symbol because the symbol was transmitted for such a long time. In order to prevent this, some wireless IoT protocols employ "data whitening." The idea is to inflict extra randomization onto the transmitted bits. A sequence of pseudo-random bits is XOR'ed with the stream of data bits. The resulting randomized bits will avoid long strings of ones or zeros. This randomization will prevent carrier recovery loops from locking onto a symbol frequency as the carrier frequency.

4.4.5 Symbol Synchronization

Much like the carrier has a frequency and phase, the symbol rate has a frequency and phase. In addition, there is ambiguity associated with symbol rate frequency and phase as between the transmitter and receiver. Symbol timing will vary between the receiver and the transmitter for the same reasons as carrier phase and frequency.

Consider the process of sampling shown in equation (4.30). The variable for time, t, is quantized into an integer number of periods defined by T_{sample}. The product of a sample index, n, and the period T_{sample} is offset by τ to determine precisely when the signal s(t) is sampled. Symbols are produced at some frequency, and the time integral of that frequency is phase. The rate of the arriving symbols determines how often symbols should be sampled. The phase of

Rectangular symbol,
no difference in amplitude
when sampling phase is offset

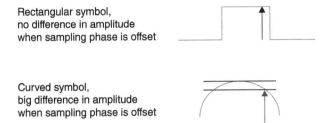

Curved symbol,
big difference in amplitude
when sampling phase is offset

Figure 4.48 Sampling in Different Symbol Shapes

the arriving symbols determines the placement of the sampling point inside a symbol

$$t = nT_{sample} + \tau;$$

$$0 \leq \tau < T_{sample};$$

$$s(t) \xrightarrow{sampling} s(nT_{sample} + \tau)$$

(4.30)

The exact point where sampling occurs in a symbol matters because of the instantaneous signal power of that sample. Instantaneous signal power is proportional to amplitude squared. In Figure 4.48, there are two shapes of symbols illustrated, a rectangular symbol and a curved symbol. Instantaneous signal power is equal for all points of the rectangular symbol. This is not the case for the curved symbol. The distance from the peak of the curved symbol to the point at which the sample is taken is shown in Figure 4.48.

Two samples along the curved symbol are compared in Figure 4.49. The sample at the center of the curved symbol in Figure 4.49 is the optimal sampling point. This is because instantaneous power is maximized. The sample at the edge of the curved symbol is of significantly reduced instantaneous signal power when compared to the sample at the center of the curved symbol.

If only the carrier had been recovered, and not symbol timing, then the signal would appear as Figure 4.50. The samples are taken at various points within the symbol. The sample period and the symbol period are different, and the received signal is sampled as it traverses between one constellation point (symbol) and another. This results in the constellation being smeared across the real-axis.

Suboptimal   Optimal

Figure 4.49 Optimal and Suboptimal Sampling Points

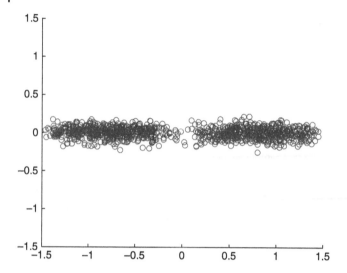

Figure 4.50 BPSK with Symbol Timing Offset

What about correcting symbol phase and not symbol rate? This question is the same as that for carrier synchronization. Correcting for the phase of the symbol will result in a temporary correction to the problem illustrated in Figure 4.50. A sample rate offset will drive the received symbol phase out of synchronization. A periodic update of the symbol phase at the receiver can be construed as a form of symbol rate correction given that such a process has a nonzero time derivative. If the symbol rate offset is small relative to the update rate with which the receiver updates its local version of the symbol phase, then a periodic phase recovery mechanism is sufficient for demodulation. Such a phase recovery mechanism can be achieved through frame synchronization. Provided that every burst from the transmitter is accompanied by a synchronization word, and that the burst is short enough for the symbol rate offset to not cause the sampling time to drift far from the optimal sampling point, extracting symbol phase information from frame synchronization can provide a sufficient means of symbol timing recovery.

For symbol-by-symbol timing recovery, a loop much like the PLL is employed. This loop is shown in Figure 4.51. The incoming signal is put through an arbitrary resampler. The output of the arbitrary resampler is tested in a Timing Error Detector (TED). The output error signal of the TED is filtered, and that filtered error drives the arbitrary resample rate.

Reference [16] gives an example and the background theory for the operation of a symbol timing recovery loop. The loop is illustrated in Figure 4.52 from [16]. The boxes labeled "polyphase MF" and "polyphase dMF" provide both a matched filter operation, polyphaser resampling, and the components for an

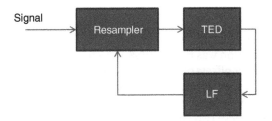

Figure 4.51 Basic Symbol Recovery Loop Structure

"Early-Prompt-Late" TED. The loop filter is a standard proportional-integral filter for control systems. The filtered error signal drives an accumulator that counts through the phases of the "polyphase MF" and "polyphase dMF" filters, causing the resample effect.

There are several techniques for designing resamplers and TEDs. These designs of resamplers and TEDs are broad subjects. This subject is far too broad to be addressed in great detail here. Instead, the reader is encouraged to read reference [17] for timing error detectors. Reference [18] addresses multirate digital signal processing and some timing error detection.

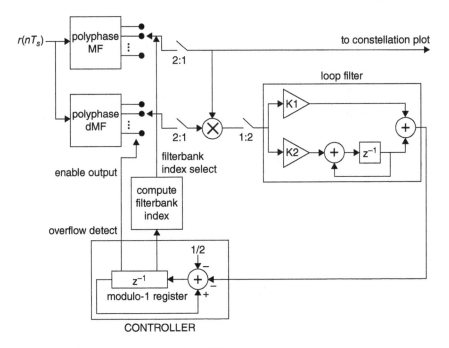

Figure 4.52 Symbol Recovery Loop from [16]

4.4.6   Order of Synchronization

Which should be applied first: Timing correction or carrier correction?

The example illustrated in this section was BPSK. BPSK was chosen for illustration because the timing and carrier offsets of BPSK are more easily separable than those in other modulation schemes. A carrier offset exclusively affects the angle of the BPSK signal whereas timing affects the measured amplitude. This allows the system designer to treat these two effects in a modular fashion.

An arctan demodulator for an FSK system also allows the designer to treat carrier and timing offsets in a modular fashion.

For other cases, the correction of timing and carrier offset can become a chicken-and-egg problem. If the carrier offset is too great, then timing cannot be estimated. If symbol timing information rests in the angle of a signal, carrier estimation becomes difficult. In such cases, it may be advantageous to utilize a two-stage approach. Use a method, such as the frame sync word, to provide a coarse estimate of timing. The sync word may also be able to provide a coarse estimate of the carrier. With these coarse estimates, the demodulator can then apply fine-tracking loops to correct for any lingering offsets.

4.5   Spread Spectrum

The goal of spread-spectrum techniques is to provide resilience in the presence of frequency-selective fading and/or interfering signals. The basic problem is illustrated in Figure 4.53 where there is a strong interferer in the bandwidth of interest. Rather than allowing such a failure on a single-frequency channel to drown out the wireless link, the spread-spectrum receiver spreads energy across multiple frequencies, allowing the link to continue operations.

Three types of spread-spectrum techniques relevant to wireless standards for the Internet of Things will be discussed here: Frequency hopping, direct sequence, and parallel sequence.

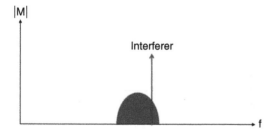

Figure 4.53 Interference within the Bandwidth of Interest

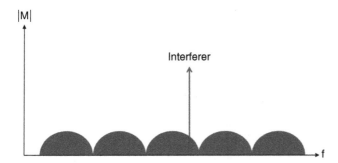

Figure 4.54 Frequency Hopping Around Interference

4.5.1   Frequency Hopping

Frequency-Hopping Spread Spectrum (FHSS) is a technique where the transmitter and receiver retune their center frequencies by way of a pre-determined sequence. Consider the problem illustrated in Figure 4.53. One way to solve the problem is to change the center frequency of the transmitted signal such that multiple frequency channels are occupied at different times. The effect of this process is illustrated in Figure 4.54. The bandwidth of the modulated signal is smaller than the total bandwidth over which that signal is frequency hopped. By way of hopping across different center frequencies, the information contained in the modulated signal can be transmitted while hopping over interference signals and across frequency-selective channels. The total power of the signal remains constant. That power is now divided among several instances of the transmitted signal across several frequency channels. Due to this, the power in any one frequency channel is reduced, as shown in Figure 4.54, as compared to Figure 4.53.

The generation process is illustrated in Figure 4.55. The figure shows a DUC where the center frequency of the NCO is driven by a sequence generator. The NCO operates as described for frequency modulation. The phase increment determines the current center frequency and can be changed for any sample period. The sequence generator contains the pseudo-random frequency-hopping sequence.

Frequency hopping allows the modulated signal to preserve a constant envelope. This makes frequency hopping an attractive method for FSK modulation schemes. The modulator in Figure 4.55 follows the FSK modulator paradigm.

The signal hops to different center frequencies in time. A hypothetical example of such frequency hops plotted against time is illustrated in Figure 4.56. This hopping provides "frequency diversity" and will allow the wireless link to persist through interference or a frequency-selective channel.

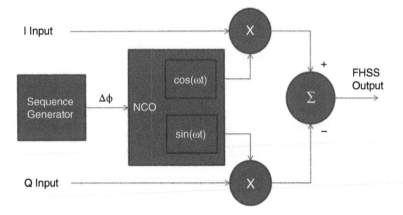

Figure 4.55 Frequency-Hopping Modulator

There is a spreading gain associated with frequency hopping. This gain is not applicable to link budgets. The signal is not made stronger in signal power, but rather the frequency diversity diminishes the effect of an interferer. The spreading gain is thus applied in the analysis of the effect of an interferer. The spreading gain for FHSS is shown in equation (4.31) to be the ratio of the hopping bandwidth divided by the modulated bandwidth.

$$FHSS\ Spreading\ Gain = \frac{Hopping\ BW}{Modulated\ BW} \qquad (4.31)$$

The receiver and transmitter need to be able to tune to the same FHSS sequence at the same time. This does take some coordination at a higher layer of processing and will be discussed in the next chapter.

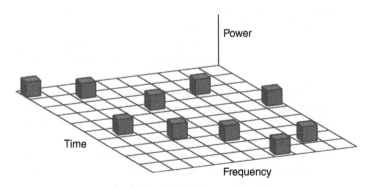

Figure 4.56 Example Frequency-Hopping Sequence

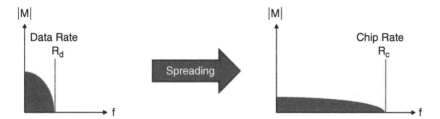

Figure 4.57 Spreading by Direct Sequence

4.5.2 Direct Sequence

Direct-sequence spread spectrum is a technique where the modulated signal is modulated again with a pseudo-random sequence of "chips." The concept of spread by way of direct sequence is illustrated in Figure 4.57. The chips have a much higher symbol rate than the original modulating signal. The effect is to spread the energy of the signal across a broader bandwidth, as defined by the "chip rate." The modulator is illustrated in Figure 4.58. This modulator follows the paradigm of the linear modulator. As is the case, DSSS is a common spread-spectrum method used for linear modulation.

The spreading process is reversible. The de-spreading process follows the modulator in Figure 4.58. When the de-spreading code is applied, the signal returns to the original non-spread bandwidth. When the signal is de-spread, the interfering signal is spread. This causes the interferer power to be dispersed across the bandwidth of the chip rate.

The total power of the signal remains constant. That power is now spread across a wider bandwidth, as defined by the chip rate. Figure 4.59 shows how the signal power is spread across the spectrum. The spread spectrum signal is smeared across a larger bandwidth. A narrowband interferer is introduced. Despreading the signal collects all the signal power into a smaller bandwidth, while causing the power of the narrowband interferer to smear across a larger

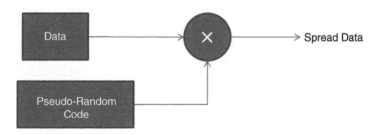

Figure 4.58 DSSS Modulator

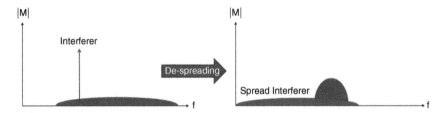

Figure 4.59 DSSS Signal with Interferer

bandwidth. The spreading gain for DSSS is shown in equation (4.32) to be the ratio of the chip rate divided by the data rate.

$$DSSS \ Spreading \ Gain = \frac{Chip \ Rate}{Data \ Rate} \qquad (4.32)$$

4.5.3   Direct-Sequence Spread Spectrum in IEEE 802.15.4

IEEE 802.15.4 uses DSSS for OQPSK and BPSK physical layers. Each has a different set of codes. All IEEE 802.15.4 networks use the same codes for DSSS. The DSSS for BPSK uses the same spreading code when transmitting a logic-high or logic-low with the polarity inverted accordingly. This process matches Figure 4.58. The OQPSK physical layer is different.

For the IEEE 8023.15.4 OQPSK physical layer, the data to be transmitted is divided into 4-bit nibbles. Those nibbles are then mapped to a 32-bit spreading code from a look-up table. The effect is a processing gain of 8 (32/4). The chips are then used to modulate the carrier, as described in Section 4.3.2.

4.5.4   Parallel Sequence

Some modes of the IEEE 802.15.4 protocol employ an Amplitude-Shift Keying (ASK) modulation scheme defined by the IEEE standard 802.15.4. This form of ASK uses a large modulation word and a form of spread spectrum called "parallel-sequence spread spectrum" [19]. Several data bits are taken in parallel, as is done for other M-Ary schemes. Each of these data bits is spread by a chip sequence. The chip sequence used on each bit is specific for each bit position in the modulation word. The resultant spread sequences are then summed together. Each chip is thus an M-Ary ASK symbol; however, it is not intended to be demodulated that way. The demodulation process utilizes circular correlation with the chipping sequences. The demodulated bits are resolved by way of spikes in the correlation output. The process is illustrated in Figure 4.60. In this example, each bit of a 5-bit modulation word is spread with a code specific to that bit position. This spreading is concurrent. The product of these

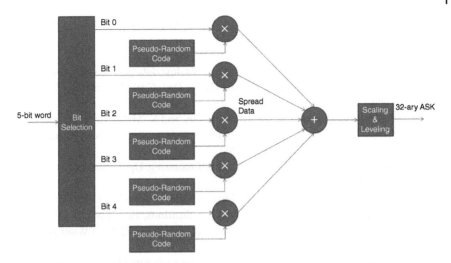

Figure 4.60 32-Ary Example

spreading operations is then summed into a combined stream. The DC component is removed from this combined stream and the amplitude is adjusted.

References

1 S. O. Rice, "Statistical properties of a sine wave plus random noise," *Bell Labs Tech. J.*, vol. 27, no. 1, pp. 109–157, 1948.

2 B. Sklar, "Rayleigh fading channels in mobile digital communications systems. Part I: Characterisation," *IEEE Commun. Mag.*, vol. 35, no. 9, pp. 136–146, 1997.

3 B. Sklar, *Digital Communications: Fundamentals and Applications*. Upper Saddle River, NJ: Prentice Hall, 2001.

4 S. Pasupathy, "Minimum shift keying: A spectrally efficient modulation," *IEEE Commun. Mag.*, vol. 17, no. 4, pp. 14–22, 1979.

5 J. H. Reed, *Software Radio: A Modern Approach to Radio Engineering*. Upper Saddle River, NJ: Prentice Hall, 2002.

6 T. Aulin and C.-E. Sundberg, "Continuous phase modulation—Part I: Full response signaling," *IEEE Trans. Commun.*, vol. 29, no. 3, pp. 196–209, 1981.

7 T. Aulin, N. Rydbeck, and C.-E. Sundberg, "Continuous phase modulation—Part II: Partial response signaling," *IEEE Trans. Commun.*, vol. 29, no. 3, pp. 210–225, 1981.

8 J. B. Anderson, T. Aulin, and C.-E. Sundberg, *Digital Phase Modulation*. New York: Plenum Press, 1986.

9 T. S. Rappaport, *Wireless Communications: Principles and Practice*. Upper Saddle River, NJ: Prentice Hall, 2002.

10 E. Lopelli, J. D. Van der Tang, and A. H. M. Van Roermund, "A FSK demodulator comparison for ultra-low power, low data-rate wireless links in ISM bands," in *IEEE Eur. Conf. Circuit Theory Des.*, Cork, Ireland, Sep. 2005, pp. II/259–II/262.

11 P. K. Meher, J. Valls, T.-B. Juang, K. Sridharan, and K. Maharatna, "50 years of CORDIC: Algorithms, architectures and applications," *IEEE Trans. Circuits Syst.*, vol. 56, no. 9, pp. 1893–1907, 2009.

12 R. G. Lyons, *Understanding Digital Signal Processing*. Upper Saddle River, NJ: Prentice Hall, 2010.

13 R. Barker, "Group synchronizing of binary digital systems," in *Communication Theory*. London: Butterworths Scientific Publications, 1953, pp. 273–287.

14 F. M. Gardner, *Phaselock Techniques*, 3rd ed. Hoboken, NJ: John Wiley & Sons, Inc., 2005.

15 J. Hamkins and M. Simon, *Autonomous Software-Defined Radio Receivers for Deep Space Applications*. Hoboken, NJ: John Wiley & Sons, 2006.

16 F. J. Harris and M. Rice, "Multirate digital filters for symbol timing synchronization in software defined radio," *IEEE J. Sel. Areas Commun.*, vol. 19, no. 12, pp. 2346–2357, 2001.

17 U. Mengali and A. N. D'Andres, *Synchronization Techniques for Digital Receivers*. New York: Plenum Press, 1997.

18 F. Harris, *Multirate Signal Processing for Communication Systems*. Prentice Hall, 2004.

19 H. Schwetlick and A. Wolf, "PSSS-parallel sequence spread spectrum a physical layer for RF communication," in *IEEE Int. Symp. Consum. Electron.*, Reading, UK, Sep.2004, pp. 262–265.

5

MAC Layer

The wireless Internet of Things (IoT) broadcasts data using the Electro-Magnetic (EM) spectrum as a medium. However, there is only one spectrum and all applications must share it. There are many different applications built upon the wireless IoT. How do those many different things mitigate contention for access to a shared medium? Consider the problem of a crowded room. A number of people congregate into a room and all start talking. In order to be heard, the people become louder to the point that no one can be heard over the din. For people, this problem might be solved by designating a coordinator or by respecting the person "holding the conch." These concepts have direct parallels for autonomous wireless devices, and we find these parallels in the Media Access Control (MAC) layer [1].

Chapter 1 presented a unified model of a stack of the lower layers of the wireless IoT. That stack is shown again in Figure 5.1. The relevant layer in the stack for this chapter is indicated by a lack of shading and an arrow. This book has progressed from the bottom-up through the stack. This chapter represents a layer of the stack that is very different from the previous two. Different responsibilities are delegated to the MAC layer as compared to the PHY or its sublayers.

Wireless IoT protocols tend to explicitly define a MAC layer. Much like the physical layer, the standards that define these layers may seem a daunting read to the uninitiated. It can help the developer of wireless IoT applications to have an appreciation as to why specific MAC layer standards were chosen for a given IoT protocol. To that end, this chapter will focus on the background theory necessary to understand the MAC layers for wireless IoT protocols. This chapter will relate those theories to actual wireless IoT examples.

The MAC layer manages access to the shared medium. The MAC layer provides synchronization between different nodes to allow wireless transmission. This synchronization becomes increasingly important as the method of access becomes more complex. As an example, synchronization may be necessary between nodes for systems employing spread spectrum. As another example, individual nodes may need permission from a controller in the wireless network

The Wireless Internet of Things: A Guide to the Lower Layers, First Edition. Daniel Chew.
© 2019 by The Institute of Electrical and Electronic Engineers, Inc. Published 2019 by John Wiley & Sons, Inc.

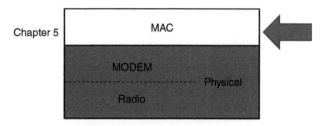

Chapter 5

Figure 5.1 Traversing the Stack: The MAC Layer

to transmit on a given wireless channel. The MAC layer manages negotiations with a control node for access.

What specific functions are encapsulated in the MAC layer varies from protocol to protocol. These functions include, but are not limited to, multiple access techniques, the synchronization of spread spectrum, and error correction. These error-checking functions may involve Forward Error Correction (FEC) and Cyclic Redundancy Checks (CRC). FEC is sometimes presented as a physical layer function. For the purposes of this book, error correction will be treated as a MAC layer function.

5.1 Bands and Spectrum Planning

There is an old adage among communication systems engineers that the most expensive part of any communications system is "spectrum." Consider these questions: At what frequencies in the EM spectrum will your communication systems transmit? What type of antenna(s) do you need? How well do signals propagate at those frequencies? And perhaps most importantly, how much will it cost to access those frequencies?

There is only one EM spectrum, and multiple services must share it. The need for some arbitration between growing numbers of wireless services was encountered in the early 20th century at the dawn of wireless services. Without any arbitration, wireless services ran the risk of colliding with one another in the use of the spectrum. Spectrum usage is, therefore, regulated and planned in most countries. Spectrum planning avoids collisions and contention in the spectrum between multiple services by allocating locations and frequencies to those services. A national regulatory authority allocates licenses to services for the use of portions of the spectrum.

The spectrum is divided into "bands" defined by the regulatory authority. Specific operators are granted license by that regulatory authority for the use of specific bands. These licenses are not free. In the United States, licensed access to these bands is sold at auctions. This cost is problematic for entrepreneurs and growing industries.

Some bands are reserved by regulatory authorities for "unlicensed" use, meaning that operators are free to access that portion of the spectrum without license, within certain operational constraints such as power and bandwidth. Operators deploying systems within these bands are expected to handle contention themselves. Different countries have allocated different bands of the spectrum for unlicensed use. There are international agreements by which regulatory authorities in different countries collaborate to set aside the same block of bandwidth for unlicensed use. Even within such a band, different countries may have slightly different rules for using that block of spectrum, but the use of that block of spectrum is generally available across borders. An example of an internationally recognized unlicensed band is the 2.4 GHz Industrial, Scientific and Medical (ISM) band. The 2.4 GHz ISM band is available for unlicensed use in most countries. This means that a single product with hardware dedicated to the 2.4 GHz band may be sold as-is across international markets. That the 2.4 GHz band is available internationally for unlicensed use makes that band an attractive choice for many wireless IoT developers. The fact that it is attractive for developers means that the band is crowded. This crowding and potential for interference is shown in Figure 5.2.

Figure 5.2 illustrates the channel mapping of Bluetooth, Bluetooth Low Energy, IEEE 802.15.4, and IEEE 802.11. The channel numbering for IEEE 802.11 channels, labeled "Wi-Fi" in Figure 5.2, is the most complex. IEEE 802.11 specifies overlapping frequency channels. This is illustrated using multiple columns in Figure 5.2. The channels for IEEE 802.11 in the 2.4 GHz band are numbered from 1 to 14; however, channel 14 is not shown in Figure 5.2. IEEE 802.11 channel 14 for the 2.4 GHz band is not permitted in some countries. Also, IEEE 802.11 channel 14 is outside the range of the other protocols shown in Figure 5.2 and does not illustrate the potential for interference. IEEE 802.11 channels 12 and 13 are not allowed in the 2.4 GHz ISM band in the United States. That means in the United States, the set of IEEE 802.11 channels 1, 6, and 11 are the largest set of non-overlapping Wi-Fi channels in the 2.4 GHz band. This makes IEEE 802.11 channels 1, 6, and 11 very popular for many network configurations. These three channels are shaded differently to highlight their importance.

The traditional Bluetooth channels, labeled "BT" in Figure 5.2, represent Bluetooth Basic Rate (BR) and Enhanced Data Rate (EDR). The traditional Bluetooth channels are numbered sequentially in the band. Each traditional Bluetooth channel has a bandwidth of 1 MHz. The channel numbers range from 0 to 79, making 80 channels in all. Bluetooth Low Energy, labeled "BLE" in Figure 5.2, takes a slightly different approach. Each of the BLE channels uses 2 MHz of bandwidth. BLE specifies channel numbers 37, 38, and 39 as "advertisement channels." Those advertisement channels are not placed in the channel plan sequentially. The placement of BLE advertisement channels is done specifically to minimize contention with IEEE 802.11 wireless local access networks

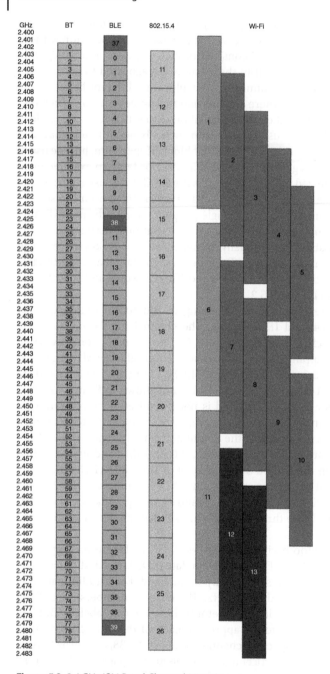

Figure 5.2 2.4 GHz ISM Band Channel Mapping

(WLAN). All other BLE channels are mapped sequentially. BLE channel numbers, including the ones not mapped sequentially, range from 0 to 39 for a total of 40 BLE channels.

Bluetooth and BLE are frequency-hopping protocols requiring a range of channels to hop across. Since WiFi exists in this range, Bluetooth and BLE will interfere with and suffer interference from Wi-Fi. Even if Wi-Fi is only consuming one of the three popular Wi-Fi channels, there is a high risk of such interference. The Bluetooth standard added an adaptive frequency-hopping technique to mitigate this possibility of interference and provide for improved co-existence between Wi-Fi and Bluetooth. This technique will be discussed in this chapter.

The IEEE 802.15.4 channels are mapped sequentially, as illustrated in Figure 5.2. Each of the IEEE 802.15.4 channels in the 2.4 GHz band uses 5 MHz. The IEEE 802.15.4 channels in the 2.4 GHz band are numbered from 11 to 26 for a total of 16 channels. Based on Figure 5.2 and the knowledge that Wi-Fi channels 1, 6, and 11 are popular, IEEE 802.15.4 channels 15, 20, 25, and 26 can most easily co-exist with Wi-Fi.

There are other unlicensed bands that are specific to individual regions. The Short Range Device band (SRD) specified in Europe is an unlicensed band that covers a bandwidth in the 800 MHz range. The United States offers an unlicensed ISM band in the 900 MHz range. In order to provide some ease and consistency in this discussion on bands, these regional unlicensed bands below 1 GHz will be referred to as the "sub-GHz unlicensed bands."

When developing an IoT application, the developer must choose a method to manage spectrum access. The very first choice is where in the spectrum the application will operate. The choice of the band of operation may restrict options in the physical layer. Regulatory authorities place different transmit power restrictions on different unlicensed bands. Such limitations on transmit power can rule out applications requiring wide areas from operating in such a band. Congestion in a band is of significant concern. There have been numerous papers written analyzing interference between different wireless networks in unlicensed bands [2, 3]. For example, Bluetooth and Wi-Fi operate in the same 2.4 GHz ISM band. Each is known to interfere with the other. While developing products for the internationally standardized 2.4 GHz band has the appeal of a large market, the congestion in that band must be addressed.

Different standards have different requirements or recommendations for the band of operation. This often depends on the application. For example, home automation can benefit from operating at sub-GHz ISM bands as those frequencies have better propagation performance in the home. ITU G.9959 focuses on home automation, and thus has operating frequencies in sub-GHz ISM bands. ITU G.9959 does not make any specific recommendations on operating frequencies because sub-GHz ISM bands are not standardized internationally. It is the Z-Wave Alliance that recommends specific operating frequencies in different countries.

The IEEE 802.15.4 protocol operates across multiple channels in several unlicensed bands. The IEEE 802.15.4 protocol specifies different physical layers for operation in those bands. These different physical layers are tailored to operate in those bands.

The system designer must provide for a sufficient bandwidth for channelization, if channelization is required. If the system designer knows that the band of operation will be highly contentious, the system designer must pay special attention to media access. Spread spectrum or adaptive systems may be deployed to address contention in the spectrum.

5.2   Spectrum Access for the Wireless IoT

A "collision" is when two nodes transmit on the same channel at the same time, with no other mitigating factor. The two transmitted signals inadvertently interfere with one another, causing a loss in the wireless link. Such an occurrence is called a "collision." There is a clear need to coordinate spectrum access across transmitters to avoid such collisions. Wireless IoT protocols employ a variety of spectrum access methods.

Channelization is a common means to mitigate contention and the chance for collisions. Channelization can take the form of frequency channels, time slots on a given frequency, or spread-spectrum techniques.

Individual nodes accessing the spectrum can be developed with the ability to intelligently move their transmissions based on observations of the spectrum. This is referred to as spectrum sensing and subsequent dynamic spectrum access. Spectrum sensing is a technique where a node in a communication system monitors a band for activity and then dynamically accesses that band. Spectrum sensing is more common in cognitive radio techniques. Reference [4] provides an overview of a proposed role for such systems in wireless IoT. Unfortunately, spectrum sensing is computationally complex and not conducive to creating low-cost devices. Therefore, a simpler mechanism called Carrier-Sense Multiple Access (CSMA) is employed. CSMA is discussed in Section 5.3.4.

For operations in the 2.4 GHz ISM band, IoT protocols can utilize a wide set of frequencies. In this band, the IoT protocol can designate numerous frequency channels and allocate those channels to individual collections of IoT nodes. However, the IoT nodes will encounter significant congestion and interference in that band. Therefore, the protocols generally specify some spread-spectrum technique.

In sub-Ghz unlicensed bands, there is less competition, but also less room for allocating frequency channels. The wireless IoT protocols operating are designed to make the most of this constrained environment. An example is the

Parallel-Sequence Spread-Spectrum (PSSS) scheme used by IEEE 802.15.4 in this band. PSSS is discussed in Chapter 4. PSSS allows the IEEE 802.15.4 protocol to enjoy some of the benefits of spread-spectrum techniques while also maintaining a relatively high spectral efficiency. Given the tight bounds on frequency usage in sub-GHz bands, PSSS provides a good compromise.

5.3 Multiple Access Techniques

Applications for the Internet of Things must choose a band in which to operate. This band dictates allowable frequencies and bandwidths. There are IoT applications that operate in environments shared by many concurrent users of that same band. The question arises: "How do multiple users transmit data concurrently across one band?" The process of allowing multiple streams across one physical resource is called "multiplexing." Dynamically allocating these limited resources to users is called "multiple access." The difference between multiple access and multiplexing is dynamic allocation.

References [1,5], and [6] provide analyses of several multiple access schemes. This section will provide a brief overview of frequency division multiple access, time division multiple access, duplexing, and CSMA.

5.3.1 Frequency Division Multiple Access

Frequency Division Multiple Access (FDMA) is a method of channelization where each user is separated by frequency. FDMA requires guard bands between channels [5]. Figure 5.3 illustrates an arbitrary FDMA system where different users of a band are separated by frequency channels. The concept of a frequency channel is discussed in Chapter 3. In Figure 5.3, there are three users at three separate frequency channels. "Guard bands" are regions between each user's bandwidth that help protect against adjacent channels' interference. The frequency channels are accessed by down converting the desired channel to baseband and low pass filtering. This down-conversion process is discussed in detail in Chapter 3.

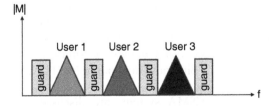

Figure 5.3 Frequency Division Multiple Access

Figure 5.4 Time Division Multiple Access

IEEE 802.11 and IEEE 802.15.4 utilize FDMA for channelization purposes. IEEE 802.15.4 uses direct-sequence spread spectrum to mitigate interference, but not for channelization.

ITU G.9959 is not dynamically channelized. ITU G.9959 does employ frequency channels, but is not aptly described as an FDMA system.

Bluetooth designates frequency channels, but Bluetooth is better described as channelized by frequency-hopping spread spectrum, as will be discussed in Section 5.4. Bluetooth Low Energy uses both frequency-hopping spread spectrum and dedicated frequency channels. Dedicated frequency channels for BLE are called "advertising channels."

5.3.2   Time Division Multiple Access

Time Division Multiple Access (TDMA) is a method of channelization where each user accesses a single frequency but at different times. Periods of time that individual users are allowed to access are called time slots. Figure 5.4 illustrates an arbitrary TDMA system where different users of a single frequency channel are separated into different time slots. The individual time slots are grouped into a TDMA frame. TDMA frames for wireless communications have guard periods to prevent inadvertent transmission collisions due to inaccuracies in timing.

TDMA systems offer several advantages over FDMA systems. TDMA systems can transmit to multiple nodes on one carrier. Having only one carrier amplified at the transmitter avoids intermodulation distortion. New time-slotted channels can be allocated more easily in a TDMA system than frequency-channels can be allocated for a FDMA system. Examples of TDMA systems can be found in second generation cellphone (2 G) protocols. For the reasons stated above, cellphone systems began employing TDMA techniques over the FDMA techniques of first generation cellphone systems.

There is a difference between "multiple access" and "duplexing." Time division duplexing will be discussed in Section 5.3.3. Bluetooth does not utilize TDMA, but rather time division duplexing.

ITU G.9959 solely employs CSMA. CSMA systems are discussed in Section 5.3.4.

IEEE 802.15.4 contains a TDMA mode for the MAC layer. Because the IEEE 802.15.4 MAC has some complexity in these optional modes, the entire subject of the IEEE 802.15.4 multiple access scheme will be addressed with "slotted CSMA" in Section 5.3.4.1.

5.3.3   Duplexing

If nodes in a wireless system are always a receiver or always a transmitter, that system is called a "simplex" [1]. A simplex system is a one-way link. A one-way pager system is an example of a simplex system.

There are also two-way links. If transmitting in a system is exclusive to one party at a time, then that system is called half-duplex [1]. Walkie-talkies are an example of half-duplex systems. Walkie-talkies are often used in a one-to-many communications link. Walkie-talkies employ privacy codes such as CTCSS to disambiguate groups of potential receivers.

If all parties in a system are free to transmit at will, then that system is called a full-duplex system [1]. An example of a full-duplex system is a phone call, where each participant can talk at will, including talking at the same time. Talking at the same time may muddle the conversation, but the wireless system will function unabated by the confusion of the participants.

Duplexing is the ability to separate two simultaneous transmissions. In order to establish a full-duplex system, both parties on the link must be able to transmit at the same time. In all of these networks and topologies, the two nodes are allowed to transmit simultaneously and must be "duplexed" to enable these concurrent streams of data. There are several methods to establish this concurrency. Among the methods of duplexing are Frequency Division Duplexing (FDD) and Time Division Duplexing (TDD).

In an FDD system, parties transmit concurrently on different frequency channels. Transmitting on different frequencies allows the transceivers of the two communicating parties to transmit at the same time, separated by frequency. FDD is illustrated in Figure 5.5. FDD is common in many wireless

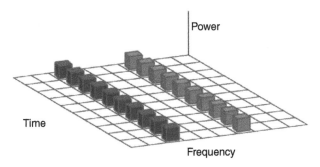

Figure 5.5 Frequency Division Duplexing

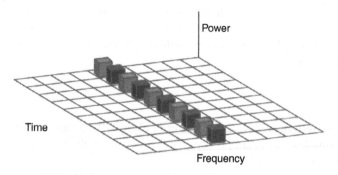

Figure 5.6 Time Division Duplexing

systems; however, it is not employed in any of the wireless IoT protocols explored in this book.

In a TDD system, both parties communicate on the same channel, but take turns using that medium. A TDD system therefore uses half-duplex links; however, it enables full-duplex communication by way of time distribution of the data. TDD is illustrated in Figure 5.6. The two parties perceive that they can communicate simultaneously even though their transceivers are taking turns.

Most of the wireless IoT protocols explored in this book do not employ duplexing. Bluetooth is the exception. Bluetooth employs TDD. The Bluetooth TDD system is illustrated in Figure 5.7, which comes from reference [7]. A Bluetooth network is an example of a star topology. Bluetooth networks are controlled by the central node called the "master" and the subscribing nodes are called "slaves." Bluetooth is a frequency-hopping system and each time slot represents a frequency hop. The frequency-hopping sequence is shared between

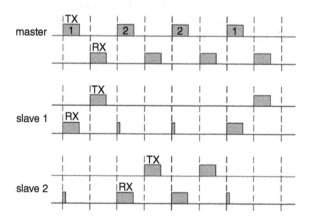

Figure 5.7 Bluetooth Time Division Duplexing [7]

the nodes. In Figure 5.7, the master node transmits on the shared frequency-hopping channel on odd-numbered timeslots. The slave nodes transmit on even-numbered timeslots.

5.3.4 Carrier-Sense Multiple Access

Consider the case where multiple nodes share a frequency channel and have no mechanism to designate and synchronize time slots. The only way to avoid collisions is to take turns accessing the singular shared channel. CSMA is a multiple access scheme, which addresses this scenario. The contended frequency channel is first "sensed" before a node transmits. This method allows a means of contention mitigation without any other synchronization or frequency usage. The problem of a singular shared medium is not exclusive to wireless protocols. Using CSMA as a solution to that problem is also not unique to wireless protocols. The wireline (not wireless) standard IEEE 802.3, also known as Ethernet, uses CSMA. Ethernet uses CSMA because there is a shared medium (interconnected Ethernet cabling) with a shared frequency channel. CSMA in wireless protocols does experience challenges unique to wireless transmissions. Those challenges will be discussed in this section.

The idea of CSMA is illustrated in Figure 5.8. There are three nodes in a network, A, B, and C. Both node A and node B want to send information to node C. Node B starts the process a little before node A does. Node B senses for energy in the channel. Sensing that the channel is clear, node B begins to transmit. Node A also senses for energy in the channel and detects the transmission from node B. Node A then waits for some amount of time, and begins the process again. Sensing that the channel is clear, node A begins to transmit. Now it is node B that senses energy in the channel and waits.

CSMA provides a means to mitigate collisions but it also creates a race condition. This is illustrated in Figure 5.9. If two nodes are about to transmit and each senses that the channel is not occupied by another signal, then both will begin transmitting. The result is a collision.

Given the possibility of this race condition, additional mitigation techniques are required. There are two common additional mitigation techniques:

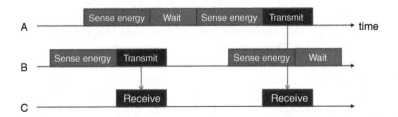

Figure 5.8 Contention Mitigation through CSMA

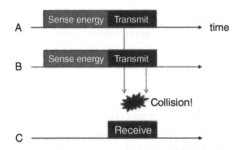

Figure 5.9 Collisions in CSMA System

Collision Detection (CSMA/CD) and Collisions Avoidance (CSMA/CA) [1,8]. In the collision detection system, the transmitting node must be able to monitor its own transmission, as it would be received at the intended node. If the transmission collides with another, then the reception is faulty. Upon detecting this, the transmitter deliberately jams the remains of the transmission, waits, and tries again. Collision detection is a solution used in Ethernet, but is not practical in a wireless system. It is not safe to assume that a transmitter can receive from all nodes from which the intended wireless recipient can receive. It is not safe to assume that the transmitter can know the delay and channel response observed by the intended wireless recipient. Repeated collisions and retries are expensive in a wireless system. Therefore, an alternative is needed for wireless systems, and that alternative is collision avoidance.

To further analyze the problem, consider the wireless network illustrated in Figure 5.10. In Figure 5.10, there is a wireless network of four nodes, A, B, C, and D. The wireless range of nodes A and B are encircled, with each node at the center of their respective ranges. Node A can transmit to and receive from node B. Node B can transmit to and receive from nodes A and C. When node A attempts to transmit to node B, node C cannot sense the transmission. Therefore, nodes A and C are bound to collide in a CSMA system. This is called the "hidden node" problem.

Another issue shown in Figure 5.10 is the "exposed node" problem. Consider what happens when node B attempts to transmit to node A. If node C attempts to transmit to node D, then node C will sense the transmission from node B to node A and delay. The link between node C and node D cannot interfere with the link between node B and node A because of the limit of the wireless range. However, because node C senses node B transmitting, node C holds off from sending data to node D. This unnecessary delay represents congestion in the network that does not exist. That is the "exposed node" problem.

Collision Avoidance is used in wireless systems as a means to mitigate the race conditions in CSMA. Collisions avoidance is found in wireless IoT standards like IEEE 802.15.4. Collision avoidance involves extra MAC layer messaging including Acknowledgement (ACK), Request-To-Send (RTS), and Clear-To-Send (CTS). The message flow diagram is shown in Figure 5.11. The sender

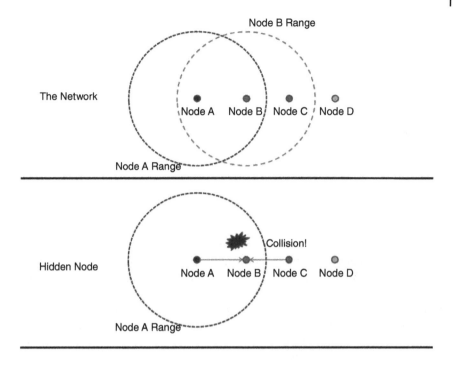

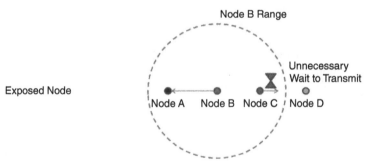

Figure 5.10 Hidden and Exposed Node

sends a "Request-To-Send" (RTS) message to the receiver. The sender then waits for a "Clear-To-Send" (CTS) message. The receiver responds with a CTS message when it is clear to send. The sender then sends the data and waits for an Acknowledgement (ACK). The receiver sends an ACK if the data arrived successfully.

This process of handshaking allows the sender to know the relevant observations at the receiver. This handshaking also provides notification to other nodes

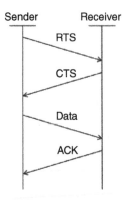

Sender   Receiver   **Figure 5.11** CSMA/CA Message Diagram

that the channel is busy or clear. Determining the busy or clear state from this handshaking is called "virtual channel sensing."

Virtual channel sensing allows nodes in the network to operate as if the channel was sensed. When one node sends a CTS message to another node, all nodes in range receive that message. The CTS message will be addressed exclusively to the node that sent the RTS message. However, all nodes in range of the CTS message will treat the channel as busy. This virtual channel sensing solves the hidden node problem in Figure 5.10. When node A needs to transmit to node B, it follows the protocol in Figure 5.11. Node C observes the CTS message being sent, and treats the channel as busy, even though node C is out of range of node A. The exposed node problem is also solved by this method. In the case of the exposed node in Figure 5.10, node B is transmitting to node A. Node C is out of range of node A and therefore does not observe the CTS message. Because node C does not observe the CTS message from node A, node C correctly treats the channel as clear and proceeds to send an RTS message to node D.

5.3.4.1 Unslotted and Slotted Systems

An unslotted CSMA system is as described in the previous section. No further coordination is required.

A slotted CSMA system confines transmission attempts to the beginning of time slots. Slotted CSMA brings the concept of time slots to the CSMA without a central coordinator dictating which transmitter may use which slot. Each transmit must contend for the use of the available time slots. Furthermore, the backoff periods are determined as an integer multiple of the timeslot period. This requires some coordination with a central node in the network.

IEEE 802.15.4 may employ slotted and unslotted CSMA systems, or a TDMA system. The IEEE 802.15.4 MAC layer has two modes: beacon enabled and non-beacon enabled. The IEEE 802.15.4 nonbeacon-enabled MAC is a CSMA system. The IEEE 802.15.4 beacon-enabled MAC can be slotted CSMA or TDMA.

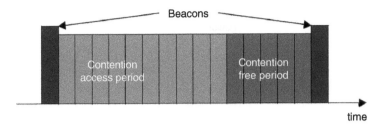

Figure 5.12 IEEE 802.15.4 Beacon-Enabled MAC [9]

The beacon-enabled frame is illustrated in Figure 5.12. Figure 5.12 comes from the IEEE 802.15.4 standard [9]. The Personal Area Network (PAN) coordinator transmits periodic beacons. Between two beacons there is a "contention access period" and a "contention free period." The "contention access period" is called such because nodes in the network are in contention to access the channel during that time. The "contention free period" prevents contention by assigning specific time slots to specific transmitters. The "contention free period" is comprised of "guaranteed time slots" allowing nodes to transmit in specific allocated timeslots, as shown in Figure 5.4. The number of slots assigned to a transmitter is variable.

5.4 Spread Spectrum as Multiple Access

Spread-spectrum techniques can provide methods of multiple access. In reference [6], this concept is referred to as "spread-spectrum multiple access."

5.4.1 Frequency Hopping

Frequency hopping depends on a sequence shared by all nodes in the link. The frequency-hopping sequence is known by both the transmitter and receiver. This a priori knowledge of the frequency-hopping sequence is necessary for the transmitter and receiver to tune to the same frequencies. This frequency-hopping sequence can be seen as a type of frequency–channel definition. Two FHSS systems may operate in the same frequency-hopping bandwidth so long as the modulated bandwidths do not overlap. In Figure 5.13, there are two signals sharing the frequency-hopping bandwidth. The two individual signals are differentiated by shading. The two signals have two different frequency-hopping sequences. These sequences may inadvertently share time-frequency pairs. This will result in a collision, as illustrated in Figure 5.13. This shows that two frequency-hopping systems can interfere with one another.

Frequency hopping covers a wide bandwidth. Due to this fact, a frequency-hopping sequence may hop into frequency channels otherwise occupied by

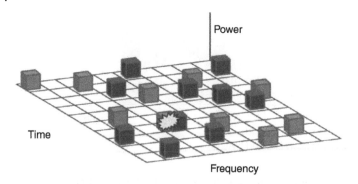

Figure 5.13 Multiple Access through Frequency Hopping

unrelated signals. For example, a Bluetooth hop sequence may employ frequencies in the 2.4 GHz band occupied by IEEE 802.15.4 or Wi-Fi. This is another source of interference for the frequency-hopping channel.

Bluetooth employs a method for collision avoidance called "Adaptive Frequency Hopping" (AFH) [7]. AFH is classified as a "non-collaborative coexistence mechanism" because it requires no collaboration with other systems using the same band. Under AFH, the Bluetooth system can sense and report whether or not reception of a transmission was possible on certain frequencies. The Bluetooth system can then build a list of frequencies to avoid. IEEE 802.15.4 and Wi-Fi use fixed-frequency channels. The usage of those frequency channels can be sensed and those channels added to a list of frequencies to avoid.

5.4.2 Code Division Multiple Access

Direct Sequence Spread Spectrum (DSSS) was introduced in Chapter 4 as a means to mitigate fading channels and interferers. DSSS uses the same technique as a multiple access scheme called Code Division Multiple Access (CDMA). CDMA is common to third generation (3 G) cellphone technologies. CDMA is a method of channelization where users share frequency and time, but are differentiated by spreading code.

IEEE 802.15.4 and Wi-Fi use DSSS; however, no wireless IoT protocol explored in this book uses CDMA. The intention of the DSSS technique employed in IEEE 802.15.4 or Wi-Fi is to mitigate interference and fading. The DSSS technique employed by those systems has no channelization capability.

5.5 Error Detection and Correction

Symbol errors occur when the symbols demodulated from a received signal are not what was transmitted by the sender. The ability of a receiver to detect these symbol errors is called "error detection." These errors can come from a variety of sources, be that some form of interference, fading channel, or noise.

Error correction is the ability to correct detected errors. Techniques employed for error detection and correction have limits as to the number and type of errors that can be detected and subsequently corrected.

Some form of error detection and correction appears in all wireless IoT protocols. Error detection and correction is a complex topic. There are entire texts dedicated to the topic, such as reference [10]. It is beyond the scope of this book to detail all the theory and variety in such a wide subject. It is, however, an important concept for all wireless systems and thus must be addressed for the wireless IoT. This section will discuss the topic and delve into concepts specific to error detection and correction as found in the standards for the wireless IoT protocols.

5.5.1 Redundancy

The goal of error detection and correction is to reduce the probability of symbol errors through redundancy. Redundancy is when information is repeated.

Repeating information in a wireless system will reduce spectral efficiency. Spectral efficiency is measured in bits per second per hertz. Spectral efficiency is the data rate divided by the bandwidth of the signal. Redundancy means that more bits will be transmitted than necessary to convey the information. Because there are more bits to transmit, spectral efficiency is reduced. Reduced spectral efficiency (bits per second per hertz) yields either an increase in bandwidth (hertz) or a reduction in throughput (bits/second).

5.5.2 IoT Considerations for Error Correction and Detection

Standards for the wireless IoT protocols employ different forms of error correction and detection. The first thing to remember when one is reading through these standards is that the authors of the standards intended for the developer to be able to meet the standards. Reading through the standards can be daunting and intimidating. It helps to demystify the standards if one considers the application for which the standards were developed. In the case of error correction and detection in the wireless IoT, consider cost, power, and latency.

Many IoT devices are low-cost devices. These devices cannot afford significant complexity. It is therefore advantageous on cost alone to avoid more complex error detection and correction schemes.

Many IoT applications cannot endure the latency incurred with complex iterative error-correction schemes. Simpler low-latency schemes are therefore preferable.

Given that power conservation is a primary concern of the IoT, the protocols are designed to make the most of the transmitted power. The protocols are therefore designed to maximize the efficiency of the transmission such that power is not wasted on unnecessary redundancy.

5.5.3 The Two Basic Types of Error Correction

There are two basic types of error correction. Those are Backward Error Correction (BEC) and Forward Error Correction (FEC). FEC allows for errors to be corrected at the receiver, whereas BEC relies upon retransmission from the sender to correct errors. FEC and BEC are not necessarily mutually exclusive and can be employed in the same wireless system.

Both BEC and FEC systems rely upon the wireless system to employ a measure of redundancy in the information transmitted.

5.5.4 Backward Error Correction

In BEC, because there is no means to correct errors at the receiver, the receiver must request a retransmission from the sender in the event that any errors are detected. In the event of detecting an error in the received packet, the receiver sends an Automatic Repeat Request (ARQ). There are numerous ARQ protocols.

In order for the receiver to detect bits errors, the sender must append some brief summary to the data bits being transmitted. This brief summary is called an error detection code. The sender does not, however, embed the means to correct detected errors. This brief summary takes the form of redundant information derived from the data bits by way of some algorithm. Upon receiving the message, the receiver runs the same algorithm on the data bits and compares the result to the received summary. If the two summaries do not match, then bit errors have occurred. Since the receiver in the BEC system cannot correct errors, the algorithm computing the summary bits can focus on maximizing the number of errors that can be detected.

5.5.5 Representing Digital Data as a Polynomial

Consider the polynomial F(x), as shown in equation (5.1). F(x) is of degree N−1. The coefficients α_1 may take a value of 0 or 1.

$$F(x) = \sum_{n=0}^{N-1} \alpha_n x^n \qquad (5.1)$$

Equation (5.1) can be used to construct a polynomial of degree N−1 to represent a binary string of length N. Example:

An 8-bit binary string with the value 10010011 can thus be represented as $F(x) = x^7 + x^4 + x^1 + 1$.

Representing digital data as a polynomial involves a concept called "Galois fields," also known as "finite fields." Specifically, what is of interest to the study of error codes in the subject of finite fields are finite fields of two elements, which are referred to as GF(2) in shorthand. The reason this is of interest is that polynomials over GF(2) allow for an arithmetic for binary strings. Treating binary strings as polynomials allows for polynomial addition, subtraction, multiplication, and division to be applied to the binary strings.

Sums and differences must be modulo 1. Therefore, addition and subtraction follow the rules for XOR, as shown in equation (5.2).

$$
\begin{aligned}
1 + 1 &= 0 \\
1 + 0 &= 1 \\
0 + 1 &= 1 \\
0 + 0 &= 0
\end{aligned}
\tag{5.2}
$$

Products follow the rules of a logical AND operation, as shown in equation (5.3).

$$
\begin{aligned}
1 * 1 &= 1 \\
1 * 0 &= 0 \\
0 * 1 &= 0 \\
0 * 0 &= 0
\end{aligned}
\tag{5.3}
$$

Multiplication of polynomials follows the standard rules for polynomials, as shown in equation (5.4), so long as equations (5.2) and (5.3) are met. In equation (5.4), the polynomial H(x) is the product of polynomials F(x) and G(x). The product of coefficients α and β follows equation (5.3). Repeating terms that result from the sum of the polynomial terms x^{n+m} follow equation (5.2).

$$
\begin{aligned}
F(x) &= \sum_{n=0}^{N-1} \alpha_n x^n; \\
G(x) &= \sum_{m=0}^{M-1} \beta_m x^m; \\
H(x) &= G(x)F(x) = \sum_{m=0}^{M-1} \sum_{n=0}^{N-1} \alpha_n \beta_m x^{n+m}
\end{aligned}
\tag{5.4}
$$

Example: The polynomial $F(x) = x^7 + x^4 + x^1 + 1$ multiplied by the polynomial $G(x) = x^1 + 1$ yields the product $H(x) = x^8 + x^5 + x^2 + x^1 + x^7 + x^4 + x^1 + 1$. The term x^1 provides a left-shifted version of F(x) to the sum, and 1 provides F(x) to the sum. The individual matching polynomial terms of H(x) can now be summed and H(x) is simplified to $H(x) = x^8 + x^7 + x^5 + x^4 + x^2 + 1$ because the two x^1 terms cancel out.

Multiplying F(x) by x^M will logical-left shift F(x) by M. The resulting product polynomial will be of order N + M-1. Concatenation can be achieved by shifting one polynomial to the left by the length of the polynomial to be appended, and then summing the shifted polynomial with the polynomial to be appended.

Example: An 8-bit binary string with the value 10010011 can thus be represented as $F(x) = x^7 + x^4 + x^1 + 1$. A second 3-bit binary string with the value 101 can be represented as $G(x) = x^2 + 1$. If the second binary string is to be appended to the end of the first to form 10010011101, the first must be left-shifted by 3 bits to form 10010011000. Logically shifting the binary string to the left by 3 bits can be represented as $F(x)*x^3 = x^{10} + x^7 + x^4 + x^3$. After $F(x)$ is shifted, the shifted version can be summed with $G(x)$ to form $F(x)*x^3 + G(x) = x^{10} + x^7 + x^4 + x^3 + x^2 + 1$.

5.5.6 Representing Bit Errors as a Polynomial

Consider the case of bit errors in a wireless system. Building upon the example above, a binary string is sent across a wireless system. The binary string can be represented as a polynomial $F(x)$ where $F(x) = x^7 + x^4 + x^1 + 1$. The receiver demodulates the wireless signal; however, there are bit errors due to noise. The receiver observes a polynomial $R(x)$, which is slightly different from the intended polynomial $F(x)$.

Errors during demodulation produce an error polynomial $E(x)$. If there were no errors, then $E(x)$ is zero. $E(x)$ is added to $F(x)$ to form the received polynomial $S(x)$, as shown in equation (5.5). $E(x)$ spans all the bits in the demodulated message, therefore $E(x)$ can be of the same degree as $F(x)$ or less. Any ones in $E(x)$ will flip coefficient values from $F(x)$, hence creating an error in $S(x)$.

$$S(x) = F(x) + E(x) \tag{5.5}$$

Now that this polynomial representation has been established, a code can be developed to help discover whether the polynomial $E(x)$ inflicted any errors on the transmitted bits $F(x)$.

5.5.7 Cyclic Redundancy Check

Cyclic Redundancy Check (CRC) is an error detection code based on the polynomial representation of binary strings and division [11].

Polynomial division, where one polynomial divides another, is much like integer division in that the operation will result in a quotient and a remainder. The division operation itself will result in a quotient, and the remainder will be ignored. A modulus operation can be used to define the remainder.

The binary data to be transmitted is represented as a polynomial $F(x)$. $F(x)$ will be of degree $N-1$. The "generator polynomial" that will be used to derive the CRC value is represented as a polynomial, $G(x)$.

Using polynomial division, a quotient, Q(x), can be defined, as shown in equation (5.6). F(x) will be of degree N–1. G(x) will be of degree M–1. F(x) is logical-left shifted by K bits. The resulting product of the shift operation is then divided by G(x). There is no fractional result in this division.

$$Q(x) = \left\lfloor \frac{F(x)x^M}{G(x)} \right\rfloor \tag{5.6}$$

The division in equation (5.6) defines the quotient, but the remainder is also needed. The remainder, R(x), can be defined using the modulus operator, as shown in equation (5.7). The remainder polynomial will be of degree M–1.

$$R(x) = \left[F(x)x^M\right] \bmod G(x) \tag{5.7}$$

The quotient and remainder relate back to the logical-left shifted product, as shown in equation (5.8). The product of the quotient and generator polynomials is summed with the remainder polynomial to produce the original logical-left shifted product.

$$F(x)x^M = Q(x)G(x) + R(x) \tag{5.8}$$

The remainder can be canceled out by adding R(x) to the logical-left shifted product. This is shown in equation (5.9). This resulting sum is perfectly divisible by the generator polynomial G(x). The sum in equation (5.9) is the message that will be transmitted to the receiver.

$$F(x)x^M + R(x) = Q(x)G(x) \tag{5.9}$$

After demodulation, the receiver has the polynomial S(x), which is the sum of the logical-left shifted F(x), the remainder R(x), and the error polynomial E(x). This sum is shown in equation (5.10). E(x) spans all the bits in the demodulated message, therefore E(x) can be of the same degree as the logical-left shifted F(x) or less.

$$S(x) = F(x)x^M + R(x) + E(x) \tag{5.10}$$

To check for bit errors, the receiver finds the remainder of the division of the demodulator output S(x) and the generator polynomial G(x). Bit-error detection D(x) is generated by this operation, as shown in equation (5.11). If S(x) is perfectly divisible by G(x), then no errors are detected. Otherwise, if the resultant D(x) is non-zero, then bit errors have been detected.

$$D(x) = S(x) \bmod G(x) \tag{5.11}$$

D(x) and E(x) may not be the same. The ability of a generating polynomial to detect bit errors depends on the roots of that generator polynomial. Generator polynomials are constructed by multiplying together different polynomial roots with different error detection capabilities. A table of those root polynomials is offered in reference [11].

5.5.8 Checksum

A "checksum" is similar to CRC in that it is an error detection code appended to the data to be transmitted. A checksum is less computationally complex as compared to CRC; however, a checksum is also far less thorough in the ability to detect errors.

There are a wide variety of checksum algorithms. The one most relevant to this discussion on standards for the wireless IoT is the checksum algorithm proscribed for data rates R1 and R2 in the ITU G.9959 standard. That checksum algorithm is called an "odd longitudinal checksum."

The binary data to be transmitted is represented as a polynomial, $F(x)$. $F(x)$ will be of degree $N-1$. $F(x)$ is restructured into a set of M-bit values. The length of the set will be an integer, L, and N divides evenly by M such that $N/M = L$. $F(x)$ will then be represented as a shifted summation of smaller words, as shown in equation (5.12).

$$F(x) = \sum_{m=0}^{M-1} F_m(x) x^M \tag{5.12}$$

The process of reformatting $F(x)$ can be visualized as reformatting $F(x)$ into a matrix of binary values. For example, a 32-bit value would be reformatted into four 8-bit words. This is illustrated in Figure 5.14. The value 0x62ECA57E, in hexadecimal, consists of four bytes (8-bit words). This four-byte value can then be reorganized into an 8×4 matrix of binary values.

A checksum value is then calculated by summing (XOR-ing) all $F_n(x)$ polynomials together, as per equation (5.2). This is shown in equation (5.15). The polynomial $C_0(x)$ is either all ones, $x^7 + x^6 + x^5 + x^4 + x^3 + x^2 + x^1 + 1$, or all zeros. This offset polynomial, $C_0(x)$, determines whether the checksum is even or odd.

$$C(x) = C_0(x) + \sum_{n=0}^{N-1} F_n(x) \tag{5.13}$$

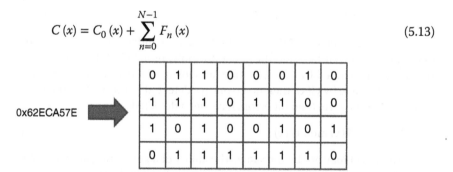

Figure 5.14 Matrix of 8-bit Words in Binary

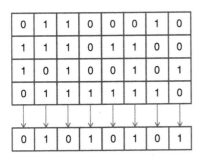

Figure 5.15 Even Longitudinal Checksum

The process of computing the checksum can be visualized as shown in Figures 5.15 and 5.16. In Figure 5.15, an even longitudinal checksum is shown. The ones in each column of the matrix are counted. If there is an even number of ones in a column, then that bit position of the checksum is set to zero. If there is an odd number of ones in a column, then that bit position of the checksum is set to one. The checksum $C(x)$ becomes the fifth row of the matrix. With the inclusion of this fifth row, the count of ones in each column is now even. This is the same as setting $C_0(x)$ in equation (5.15) to all zeros.

Figure 5.16 shows the "odd" version of this process. As before, the ones of the columns of the matrix are counted. This time an odd number of ones is desired. The ones in each column of the matrix are counted. If there is an even number of ones in a column, then that bit position of the checksum is set to one. If there is an odd number of ones in a column, then that bit position of the checksum is set to zero. The checksum $C(x)$ becomes the fifth row of the matrix. With the inclusion of this fifth row, the count of ones in each column is now odd. This is the same as setting $C_0(x)$ in equation (5.15) to all ones, $x^7 + x^6 + x^5 + x^4 + x^3 + x^2 + x^1 + 1$. The checksum algorithm proscribed for data rates R1 and R2 in the ITU G.9959 standard is called the "odd longitudinal checksum."

The placement of bit errors is very important to whether or not the checksum can detect them. This problem is illustrated in Figure 5.17. In Figure 5.17, the same data is transmitted in the matrix on the left and the matrix on the right. Two bit errors are injected in both cases. For the case on the left, the two bit

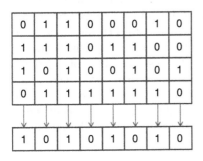

Figure 5.16 Odd Longitudinal Checksum

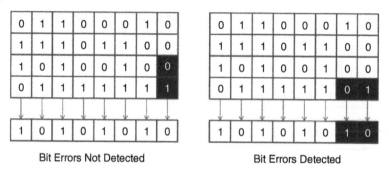

Bit Errors Not Detected   Bit Errors Detected

Figure 5.17 Detecting Bit Errors in Odd Longitudinal Checksum

errors occur in the same column, resulting in the calculation of the expected parity bit. This fails to detect errors. For the case on the right, the two bit errors occur in two different columns. This results in two incorrect calculations of parity bits, and both errors are detected. The longitudinal checksum, whether odd or even, can only detect one error in each column. If any even number of bit errors arrive in the same column, the errors cancel out the effect on the checksum. It cannot be said how many bit errors the checksum can detect because the ability to detect those errors depends upon placement.

5.5.9   Forward Error Correction

In Forward Error Correction (FEC), the sender embeds the means to both detect and correct errors into the transmitted signal. This is called an error correction code. This process converts a sequence of data bits into a larger sequence of coded bits to be sent over the air. It is important to clearly define the term "coded bits." These coded bits are directly derived from the original data bits. A resultant string of coded bits will be longer than the original string of data bits. If the original data bits are included unmodified in the output string of coded bits, then that type of FEC code is called a "systematic" code. This definition allows for a generalization of various FEC methods.

Among the simplest method is to repeat the original data bits an odd number of times. This method clearly incorporates the unmodified data bits into a longer string of what are now defined as "coded bits." Some FEC methods create a string of coded bits from a string of data bits by appending a smaller sequence of derived bits to the end of blocks of data bits. The result is a string of coded bits that include the unmodified original bits and is longer than the string of original data bits. This method is like the cyclic redundancy check, except that some errors detected can be corrected. Some FEC methods derive a string of coded bits by employing convolutional techniques. The string of coded bits resulting from such a process will be longer than the original string of data bits. Such a process may or may not include the unmodified original data bits in the resulting string of coded bits.

This larger sequence of coded bits is derived from the data bits. The coded bits contain no new information themselves. This process of converting data bits to a longer sequence of coded bits deliberately adds redundancy to the transmitted information. The receiver decodes the coded bits, and uses that code to hopefully detect and correct any bit errors that have occurred. The FEC coding must be designed with the ability to both detect and correct errors. The ability to detect and correct bit errors depends upon the amount of redundancy added and how that redundancy has been added.

5.5.9.1 Redundancy in FEC

The amount of redundancy in a forward error correction code is measured in "coding rate," r. The definition of r is provided in equation (5.14).

$$r = \frac{N}{K} \tag{5.14}$$

where r is the code rate, N is the number of data bits, and K is the total number coded bits. The number of coded bits is always greater than the number of data bits. Therefore, the code rate, r, is always less than 1.

The coding process transforms the data bits, $F(x)$, into coded bits, $C(x)$. $F(x)$ will be of degree $N-1$, and $C(x)$ will be of degree $K-1$. K is greater than L, and this represents the redundancy. The sender will send the coded bits to the receiver. Some types of FEC add a summary of length M to a string of data bits of length L, where $K = L + M$. This type of FEC is called "block codes" because a block of data is encoded. Appending this summary to the end of the string of data bits is similar to the process used for CRC. The difference between CRC and block codes, functionally speaking, is that the redundancy in block codes is designed to identify and locate errors in the encoded string.

The measure of spectral efficiency (bits per second per hertz) for a system is determined by the number of data bits transmitted, not the number of coded bits transmitted. Therefore, a lower code rate means less spectral efficiency.

Example: A system uses a $\frac{1}{2}$ code rate and transmits 1 Mbps of coded data over a 2 MHz bandwidth. Because the system uses an FEC code with a rate of $\frac{1}{2}$, the actual data rate is 0.5 Mbps. The spectral efficiency, thus, is calculated to be 0.25 bps/Hz. The data rate of the system is 0.5 Mbps. If the system transmitted uncoded data, then the data rate would be 1 Mbps and spectral efficiency would be 0.5 bps/Hz. The $\frac{1}{2}$ FEC code cuts spectral efficiency in half.

The idea is to get the code rate as close to unity as possible while maximizing the ability to detect and correct errors.

5.5.9.2 Hamming Distance

The receiver receives S(x), which is the transmitted C(x) plus an error polynomial E(x). As the receiver reverses the coding process, it may be able to detect and correct errors inflicted on the transmitted signal. To do this, a concept of "Hamming distance" is employed. Hamming distance is a measure of bitwise difference between two binary strings.

The ability to detect and correct errors is related to the distance between the strings being used. Error correction codes can detect and correct some number of bit errors, as determined by the minimum Hamming distance. The minimum Hamming distance the smallest Hamming distance between any binary strings sent by the transmitter.

The number of bit errors that can be detected by an error correction code, D, is given by equation (5.15). The number of bit errors that can be detected is a function of the minimum Hamming distance, d.

$$D = d - 1 \tag{5.15}$$

The number of bit errors that can be corrected by an error correction code is given by equation (5.16). C is the number of bit errors that can be corrected. The value is floored to the nearest integer.

$$C = \left\lfloor \frac{d-1}{2} \right\rfloor \tag{5.16}$$

5.5.9.3 (3,1) Repetition Code

An easily illustrated example of redundancy is to repeat the transmitted information three times. The receiver estimates the symbol value three times, and the majority vote wins. This is called a "(3,1) repetition code." This name uses the formatting (n,k), where n is the total number of bits and k is the original information bits. The resultant redundancy gives the receiver three opportunities for symbol decision.

Redundancy comes at the expense of reducing data throughput. In the example of the (3,1) repetition code, the throughput is reduced to a third, but has a better chance of being successfully received in a noisy environment.

The minimum Hamming distance is illustrated in Figure 5.18. Only two states are shown, 111 and 000. These are the only two states transmitted.

The minimum Hamming distance of the (3,1) repetition code is 3; therefore, the code can detect up to two bit errors and correct up to one bit error. This Hamming distance of 3 comes at the expense of a significant reduction of data throughput to 1/3 the rate that would be enjoyed if the data were transmitted

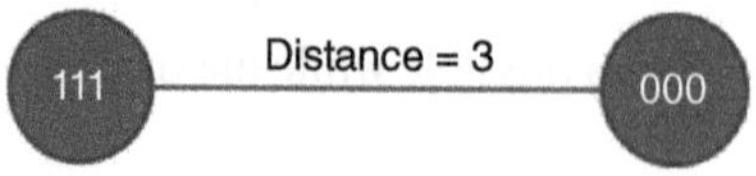

Figure 5.18 (3,1) Repetition Code Distance

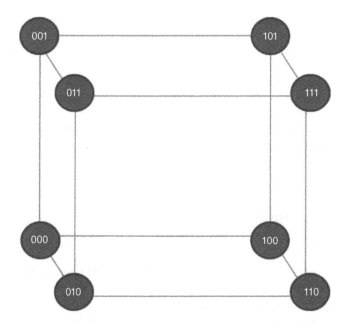

Figure 5.19 States of the Received Coded Bits

uncoded. Therefore, the (3,1) repetition code is very inefficient as compared to other FEC coding schemes, which can achieve the same Hamming distance with more spectral efficiency. The main benefit of the (3,1) repetition code is simplicity.

In an FEC system, the receiver will attempt to detect and correct errors. The receiver will receive the original coded message $C(x)$ summed with an error vector $E(x)$. For the example of the (3,1) Repetition Code, this means that the receiver will receive bits that can be plotted in the three-dimensional space shown in Figure 5.19. Each bit is a represented as a dimension. Logical low, 0, is represented as being the opposite of logical high. The only two valid positions in this space are (1,1,1) and (0,0,0) as these are the only two strings that could be transmitted. All other locations are the result of the error vector, $E(x)$, being summed with the transmitted message $C(x)$. The receiver chooses the valid state with the shortest Hamming distance to the received state.

A sequential implementation of the (3,1) repetition code is to see each bit repeated three times in a row, as shown in Figure 5.20. The repeated bits are

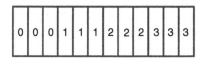

Figure 5.20 (3,1) Repetition Code, Implemented Sequentially

placed sequentially such that each bit is repeated three time before the next one begins.

When placed sequentially, the bits can be seen not as a string of coded bits but rather as the original data bits transmitted at a lower data rate. The sequential (3,1) repetition code, as illustrated in Figure 5.20, can therefore be used as reduction in the data rate. This will boost the energy per bit when demodulated.

If the channel has no small-scale fading, then such a sequential implementation can be employed to give a boost to important parts of a data packet, such as the header. However, this sequential approach may be problematic in more complex channels. Such a discussion goes beyond the scope of this book. The reader is encouraged to explore reference [5].

5.5.10 FEC and BEC Comparison

All wireless IoT protocols use BEC, but few use FEC. Why would this be the case?

BEC can detect more errors with less redundancy than FEC. BEC allows for greater error detection than FEC. When a BEC method detects an error, the transmitted message will be repeated. Therefore, it is advantageous to employ a BEC method in the wireless system to better guarantee the delivery of an error-free payload.

FEC allows the signal the possibility of surviving bit errors without the need for retransmission. This sounds better on the surface. The problem is that FEC has a higher up-front cost than BEC in terms of spectral efficiency. Whether or not bit errors occur with any significant frequency, the redundancy required for FEC will reduce throughput as compared to a BEC-only system. For wireless systems that expect a low probability of bit error, the BEC-only method is optimal.

Systems that employ FEC perform better than BEC-only systems as the probability of bit error increases. A system that encounters a high enough probability of bit error can actually achieve higher throughput when FEC is employed even with that constant redundancy. Consider that as the possibility of bit error increases, the number of repeated transmissions in a BEC-only system will increase. Increased retransmission means a reduced throughput. The FEC system can correct bit errors, and thus suffer fewer retransmissions.

5.5.10.1 Bluetooth and FEC
Bluetooth uses the (3,1) repetition code as the FEC code for the header of Bluetooth packets. The (3,1) repetition code is implemented with the bits transmitted sequentially. Each header bit is repeated three times in a row, as shown in Figure 5.20. The sequential (3,1) repetition code, as employed in Bluetooth headers and illustrated in Figure 5.20, can therefore be used as reduction in the data rate and boost the energy per bit when demodulated.

Bluetooth Basic Rate (BR) and Enhanced Data Rate (EDR) use both BEC and FEC. The Bluetooth header uses the (3,1) repetition code, and the Bluetooth packet is encoded with a "(15,10) Hamming code," which is a 2/3 FEC. Hamming codes are block codes and function, as described in this chapter. A (15,10) code will be difficult to illustrate and visualize in contrast to the (3,1) code. More information on Hamming codes can be found in reference [10].

Bluetooth Low Energy (LE), which was designed as a wireless IoT addition to the Bluetooth standard, does not mandate the use of FEC. Bluetooth LE offers various coding rates, as high as r = 1/8. Bluetooth LE, much like many other wireless IoT protocols, relied upon error detection rather than error correction. FEC was made optional in a later version of Bluetooth LE for the purpose of extending range at the expense of data rate.

5.6   Energy Efficiency

How long will your battery last in a battery-powered wireless transceiver? This is an essential question to answer. What will determine the answer to that question are the power consumption and energy efficiency of the wireless system. Power consumption is the number of joules consumed over time. Energy efficiency is the number of joules consumed per bit of throughput.

Chapter 4 discussed energy efficiency in selecting modulation schemes. Frequency-shift keying allows for nonlinear amplifiers and therefore makes the transmit chain more efficient. However, the management of how much energy a wireless system consumes does not stop at the transmitter. Requiring the receiver to be on and consuming power costs energy. It is not the physical layer, but rather the MAC layer that will determine power consumption and the energy efficiency of the wireless system.

Sections 5.5.10 and 5.5.10.1 compared the costs of FEC and BEC in terms of transmit time and retries. All transmission times cost energy. Every redundant bit costs energy. Redundancy costs extra energy, though that redundancy may save energy in terms of decreasing the need for retransmissions if bit error rates are high. As an example of redundancy costing energy, Bluetooth Core Specification version 5.0 [12] contains an optional FEC for Bluetooth LE but the specification goes on to warn that the user carefully consider the impact of the coding on power consumption. Consider that the optional FEC may be as high as r = 1/8, meaning that there are 8 code bits for every one data bit. The time needed to transmit such a payload of code bits is eight times longer than that needed to transmit the data bits. The transmitter and receiver will be consuming at least eight times as much energy. That fact should make clear that redundancy costs energy. Retransmission also costs power. Every time a transmission must be repeated, the cost in energy will at least be doubled. If an uncoded transmission takes multiple tries to be received without a CRC failure,

Power
Being Consumed

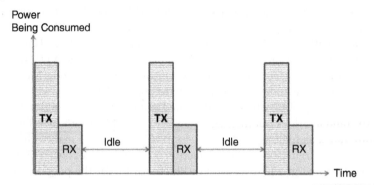

Figure 5.21 Generic Transceiver Sleep Mode Duty Cycle

then that too dramatically hurts energy efficiency. Therefore, the FEC coded variant of Bluetooth LE is effective for longer distances, but at shorter distances, it is inefficient in terms of energy and battery life.

The MAC layer is in charge of putting idle nodes in the network to sleep in order to save battery life and reduce unnecessary power consumption. The MAC layers of the wireless IoT are therefore designed such that battery-dependent nodes need not continuously monitor for transmissions. ITU G.9959 specifically states that nodes can spend most of their time in sleep mode [13]. Sleep mode refers to neither receiving nor transmitting. IEEE 802.15.4 and the Bluetooth Core Specifications for Bluetooth LE also contain sections stating the concern for conserving power. A common solution to the concern over power consumption is to implement a duty cycle. Figure 5.21 illustrates the concept. Some time is spent transmitting (TX), some time is spent receiving (RX), and as much time as possible is spent idle. The cycle is periodic. Transmit consumes the most power. Receive consumes the second most. Power consumed over time is energy spent on that task. The concept of an inactive period allows for a duty cycle.

This duty cycle must be timed across different nodes. Those nodes may not share a timing reference. That means the synchronization will skew over time. This problem is illustrated in Figure 5.22. The clocks in nodes 1 and 2 in Figure 5.22 are slightly different. Due to this slight variance, the duty cycle slips and the two are no longer aligned.

There are several solutions to this problem of synchronizing duty cycles. Section 5.3.4.1 discussed beacon-enabled frames of IEEE 802.15.4. Those frames have the option of an "inactive" period where more power-sensitive nodes can shut down. This is illustrated in Figure 5.23 [9]. Data is transferred within the active period, and then all nodes beholden to that beacon-enabled frame go idle until the next beacon. The nodes operate independently but can keep timing synchronized by way of the beacons.

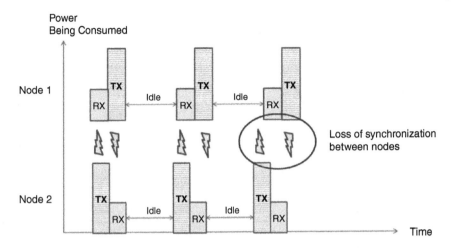

Figure 5.22 Generic Transceiver Duty Cycle, Nodes Out Of Sync

ITU G.9959 takes a different approach. The nodes in ITU G.9959 are either "always listening" or "frequently listening" (FL). The FL nodes have a "wakeup interval." The FL nodes use the wakeup interval as a timer to determine when to come out of sleep mode and listen for activity [13]. These duty cycles are run independently in each node. A node attempting to contact a sleeping node will transmit a special set of frames called "beam frames" for a length of time longer than the sleep period. This elongated transmit period serves two purposes. It reserves the channel for the pending data transfer and it guarantees that the

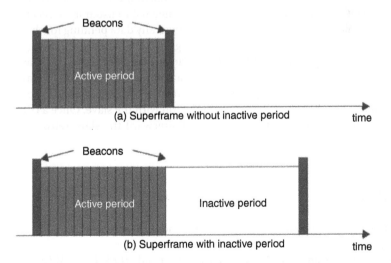

Figure 5.23 Inactive Periods in Beacon-Enabled IEEE 802.15.4 Frame [9]

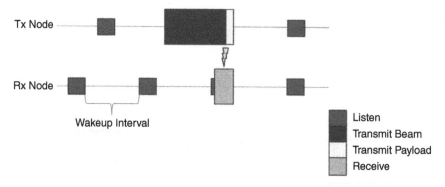

Figure 5.24 ITU G.9959 Beam and Wakeup Intervals

sleeping recipients will eventually awaken to see the beam frames. A message is transmitted after the beam frames. This concept is illustrated in Figure 5.24. The Tx node and the Rx node spend most of the time in sleep mode. The Tx node has some data to transmit to the Rx node, so the Tx node begins to transmit beam frames. The Tx node does not return to sleep until after the transmit operation is complete. The wakeup interval of the Rx node expires and the Rx node receives the beam frames. The Rx node does not return to sleep mode because there is a pending message. The Tx node transmits the message at the end of the beam frames. Then both the Rx node and the Tx node return to sleep mode.

Bluetooth LE also addresses reducing power consumption. In the original Bluetooth specification, slave nodes in a piconet needed to listen in all receive periods to the master node to see whether there was any data pending for them [12]. This concept is shown in Section 5.3.3 in Figure 5.7. Bluetooth LE changed this paradigm such that the master listens to advertising channels. The slave nodes spend most of their time asleep. When a connection is desired, a slave node pages the master on the advertising channels in order to request a connection. The master then coordinates a connection with the slave. Once a connection is established, data is transferred, and afterward the slave returns to sleep.

References

1 A. S. Tanenbaum, *Computer Networks*. Upper Saddle River: Prentice Hall, 2003.
2 I. Howitt, "WLAN and WPAN coexistence in UL band," *IEEE Trans. Veh. Technol.*, vol. 50, no. 4, pp. 1114–1124, 2001.

3 S.-H. Lee, H.-S. Kim, and Y.-H. Lee, "Mitigation of co-channel interference in Bluetooth piconets," *IEEE Trans.Wireless Commun.*, vol. 11, no. 4, pp. 1249–1254, 2012.

4 A. A. Khan, M. H. Rehmani, and A. Rachedi, "Cognitive-radio-based Internet of Things: Applications, architectures, spectrum related functionalities, and future research directions," *IEEE Wireless Commun.*, vol. 24, no. 3, pp. 17–25, Jun. 2017.

5 B. Sklar, *Digital Communications: Fundamentals and Applications*. Upper Saddle River, NJ: Prentice Hall, 2001.

6 T. S. Rappaport, *Wireless Communications: Principles and Practice*. Upper Saddle River, NJ: Prentice Hall, 2002.

7 Part 15.1: Wireless Medium Access Control (MAC) and Physical Layer (PHY) Specifications for Wireless Personal Area Networks (WPANs), IEEE 802.15.1-2005, 2005.

8 J. Burbank, W. Kasch, and J. Ward, *Network Modeling and Simulation for the Practicing Engineer*. Hoboken, NJ: Wiley-IEEE, 2011.

9 Part 15.4: Low-Rate Wiress Networks, IEEE 802.15.4-2015, 2015.

10 S. Lin and D. J. Costello, *Error Control Coding: Fundamentals and Applications*. Prentice Hall, 1983.

11 W. W. Peterson and D. T. Brown, "Cyclic codes for error detection," *Proc. IRE*, vol. 49, no. 1, pp. 228–235, 1961.

12 Bluetooth Core Specification version 5.0, Bluetooth Special Interest Group, 2016.

13 Recommendation ITU-T G.9959 Short range narrow-band digital radio communication transceivers – PHY, MAC, SAR and LLC layer specifications, International Telecommunication Union, 2015.

6

Conclusion

Chapter 1 presented a unified model to describe a protocol stack for the wireless Internet of Things (IoT). That model is shown in Figure 6.1. The model decomposes the functions of such a wireless IoT system into layers. This book focuses on the lower layers that are defined by open standards.

Chapter 2 presented general information on the standards covered in this book in order to provide context to the subsequent chapters. Those subsequent chapters traversed the stack shown in Figure 6.1 from the bottom-up, reviewing background theory and relating that theory to specifications in relevant standards. The intention of this process was to help the uninitiated navigate the language of the standards, to identify common threads in the standards, and ultimately to elucidate those standards.

Here, at the conclusion of that traversal, this book offers a few final thoughts on selecting the standard appropriate for your needs and the growing standardization in the upper layers in the wireless IoT.

6.1 Selecting the Right Standard

How does one select the wireless Internet of Things (IoT) standards appropriate for a given application? Unfortunately, there is no easy answer. The standards explored in this book have significant overlap in the types of applications that can be satisfied by that standard. The primary goal of this book was to develop within the reader an understanding of the nuance involved when comparing these wireless standards. Here are some factors one may consider when comparing standards.

6.1.1 Cost

The more complex a system, the more it costs. Simplicity may not meet the optimal points for other considerations, but it has a great chance of minimizing cost. Chapter 4 discusses inexpensive approaches to MODEM development.

The Wireless Internet of Things: A Guide to the Lower Layers, First Edition. Daniel Chew.
© 2019 by The Institute of Electrical and Electronic Engineers, Inc. Published 2019 by John Wiley & Sons, Inc.

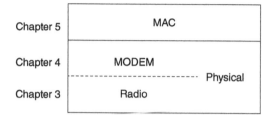

Figure 6.1 Unified Protocol Stack, Lower Layers

6.1.2 Data Rate

How often will the nodes in the system be transmitting and how much data will those nodes need to transmit? The per packet data rate and frequency of transmitting packets based upon your system needs will be the primary consideration when determining which technology to use. Chapter 2 lists data rates supported by the standards discussed in this book. Chapter 4 discussed how to synchronize with those rates. Please note that the more the synchronization required, the more complex the system, and the more expensive that system will be.

6.1.3 Band of Operation and Environment

The developer must consider their environment. Is the system to be used indoors or outdoors? What channel can the developer expect? The developer can refer to the link budget calculations in Chapter 3 to see the effects of the environment and the expected range for a given transmit power. The selection of a band of operation also plays a role in range calculation. Bands of operation are discussed in Chapter 5.

6.1.4 Network Topology

How many nodes will be in your network? The different standards discussed handle that problem differently. Some can handle many more nodes. How much throughput do you need? The standards discussed have offered multiple modes with different data rates. Therefore, one cannot necessarily discount a standard given a data rate. There may be significant overlap for the desired data rate. Network topology was discussed in Chapters 1 and 2. This book focuses on the lower layers. Network topology is important to a system, but is at the edge of the scope of this book.

6.1.5 Energy Efficiency

What about batteries? A high transmit power will be a drain on a battery. Inefficient linear amplifiers will also be a drain on battery life. If battery life is important, then it will be beneficial to use a constant envelope modulation scheme, as discussed in Chapter 4. Chapter 4 discusses inexpensive demodulation techniques. Consideration of energy efficiency does not end at the physical layer. In fact, the system must rely on the MAC layer to expend as little energy as possible in order to meet the objectives of the application. Chapter 5 discussed the MAC layer's role in minimizing the energy used in a wireless system.

6.2   Higher Layer Standardization and the Future of IoT

An active area of development, and a key component in the future of the wireless IoT, is standardization in the upper layers. This book has expounded on the standards that provide interoperability in the lower layers of a wireless IoT protocol stack. This interoperability allows hardware from different vendors to join together to form a wireless IoT network in support of various applications. But what about the upper layers?

The protocol stack from Chapter 1 has been expanded in Figure 6.2 to include additional upper layers so as to provide context to the subsequent discussion. The shaded layers in Figure 6.2 are the "upper layers" not covered by the lower layer standards discussed in the previous chapters.

This book has traversed up the stack, starting at the most physical interface to the medium, to control over access to that medium. As one traverses further up the stack, the requirements of the layers become more application-specific. The upper layers must address the needs of the end-user application. Should the end user expect interoperability in those upper layers?

The network layer is an example where such standardization is taking place. Chapter 2 discussed changes in ITU G.9959 from the 2012 standard to the 2015 standard. This transition added logical link control as a means to allow network implementations other than the one defined by the Z-Wave Alliance. The 2015

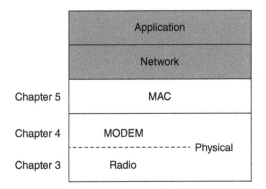

Figure 6.2 Unified Protocol Stack, Lower Layers

version of ITU G.9959 specifically names 6LoWPAN as such a network layer implementation. Chapter 2 also discussed the inclusion of a mesh topology feature in Bluetooth Low Energy, outside of Bluetooth Core Specifications 5.0.

Without interoperability in the upper layers, every vendor would be free to require a different interface, a different graphical user-interface program, a different phone app, to interface to the network formed by the standardized lower layers. It is not unreasonable that a user should expect interoperability in the upper layers. Standardization at the application layer is of great importance to the wireless IoT and the end user. Many industry groups are developing such standards for their own protocol of interest, such as "application profiles." Such a profile imposes some rules for interoperability on different manufacturers and vendors of products for a given application. Independent standards bodies are also working to standardize the upper layers across industry groups. Ultimately, such upper-layer standardization is important to the future of the wireless IoT and is an area of exciting new development.

Index

The Wireless Internet of Things: A Guide to the Lower Layers, First Edition. Daniel Chew.
© 2019 by The Institute of Electrical and Electronic Engineers, Inc. Published 2019 by John Wiley & Sons, Inc.